Demonstrations and Activities

for

HIGH SCHOOL CHEMISTRY

Demonstrations and Activities

for

HIGH SCHOOL CHEMISTRY

John H. Woodburn
Lewis R. Sanford
W. James Brawley, Jr.

Parker Publishing Company, Inc. *West Nyack, N.Y.*

LIBRARY OF CONGRESS
CATALOG CARD NUMBER: 73-152442

PRINTED IN THE UNITED STATES OF AMERICA

ISBN-0-13-198358-X
B&P

DEDICATION

To our lab assistants, those young men and women who wash the glassware, get equipment out and put it away, run errands, buffer our impatience, take the blue sky out of some of our ideas yet inspire us to have new ideas, and restore our perspective and enthusiasm when the frustrations of being teachers exceed, just temporarily, the satisfactions.

HOW TO USE THIS TEACHING TOOL

This book is organized around the major topics of high school chemistry. Each of the chapters opens with a brief discussion of what our students may well be expected to know or to be able to do when they come to us and when they leave. Obviously, in a book of this size and nature, the discussion is in terms of goals; the authors do not attempt to condense or restate the principles and concepts of chemistry.

EMPHASIS ON REACTIONS

Students seem to learn chemistry most effectively when instruction provides opportunities for them to actually see, smell, and sometimes, feel or hear the direct evidence of chemical phenomena. On this basis, the authors rely heavily on instructional activities built around chemical reactions. They fall back upon the literature of our profession and their own classroom experience to find the reactions which best confront students with the observations most likely to lead our students to grasp and appreciate the big ideas, the principles and concepts of chemistry.

Among the thousands of chemical reactions known or contemplated, those reactions have been selected for major emphasis which support a more or less standard introductory course. Additional reactions are suggested, however, whereby teachers can revise, supplement, or extend their repertories of "tried and true" activities. As best they can, the authors call attention to safety precautions, especially when suggesting the use of chemicals not encountered ordinarily in the usual undergraduate chemistry courses.

PROCEDURES ARE PRESENTED INDIRECTLY

Most chemistry teachers soon develop their individual teaching methods, tactics, and procedures. The best way to catalyze this development seems to be to provide opportunities to adapt or modify the procedures other teachers have found to be effective. Following this argument, the authors include in the early chapters detailed descriptions of their way of handling selected activities. The general pattern of the activity can be lifted from the illustration and the reader can then adapt the method to his own way of doing things. As the book unfolds, the instructional aspects of each activity tend to be presented with increasingly less detailed treatment. The goal is to help each teacher accumulate a wide enough repertory of methods and procedures to allow him to match his treatment of the day's topic with an instructional format that fits the characteristics of his students.

DISCOVERY, INQUIRY, REINFORCEMENT, OR DRILL

We all know how satisfying it is to have our students think through their own observations of a chemical event, to "discover" orderly relationships, and then to identify a basic principle or put together a major segment of a chemical concept. Even more rewarding are those instances when our students are captured by a curious aspect of a reaction and launch their own investigations in pursuit of its comprehension.

Unfortunately, the capabilities required to bring about the "discovery" type of learning are not always innate. More often, these capabilities must be patiently and astutely evoked and nurtured. Hence, although many of the activities described in this book may seem to rely heavily on the presentation and comprehension of information, underlying these activities is the goal of readying the student increasingly for investigative and discovery capabilities and attitudes.

PROBLEMS OF CLASSROOM AND LABORATORY MANAGEMENT AND DISCIPLINE

Ways to avoid or solve discipline problems seem to depend very much on the personal traits and mannerisms of individual teachers. On this basis, solutions are more appropriately suggested than prescribed. Thus, between the lines, so to speak, of their detailed descriptions of instructional activities, the authors suggest ways to arouse or maintain student interest; create a general climate that promises to avoid or diminish discipline problems; reduce clutter, confusion, and breakage in the laboratory; diminish rather than enhance opportunities and urges for cheating and dishonesty; and neutralize similar routine but highly significant worries and aggravations.

THE PROBLEMS OF INDIVIDUAL DIFFERENCES

In many schools, students come to their chemistry course with wide differences in reading skills and other study techniques; in the ability to do quantitative thinking and computing comfortably; in their grasp of the elementary principles of science and the ability to fit them into new situations; in the ability to use the laboratory effectively; in taking part in and benefiting from group instruction; and many other known and unknown factors which seem to influence student success or failure. These widely ranging differences put a teacher in an age-old dilemma. To set goals too low risks boring the well prepared and ambitious student. Goals set too high risk creating gaps between where his students are and where instruction begins with subsequent discouragement and frustration for all concerned.

Actually, the goals of high school chemistry are pretty much set by the nature of the course and what lies ahead for its students. Each of us feels the urge to meet these goals. This requires us, with as much finesse as possible, to overcome the weaknesses of our inadequately prepared students by reviewing or reteaching essential albeit supposedly familiar material or skills. At the same time, we must, at best, avoid holding back our best students and, hopefully, keep them working toward goals appropriate for their abilities and interests. This is the primary purpose behind much of the material we include in the sections titled, "Ideas for Additional Student and Teacher Activities."

THE PROBLEM OF LINEAR VERSUS SPIRAL COURSE ORGANIZATION

The high school chemistry course scarcely lends itself to a linear organization. Usually, we introduce our students to certain principles or concepts, take them so far, and then turn to other closely integrated principles or concepts. Sooner or later we return to the ideas that were introduced some time ago in order to add to the precision or to enhance appreciation. This can be a source of serious frustration to students (or teachers)who have taken pride in "complete" mastery of whatever lesson was to be learned, at least mastery within the limits set by the teacher or course of study.

The reader can well visit and revisit each chapter of this book as he moves his students through the usual sequence of exposure and re-exposures to the principles and concepts which make up the high school chemistry course. Not only is this consistent with the spiral concept of course structure but it also allows us to "get in a few licks" on related material while paying primary attention to the chief target for the day. Furthermore, chemical phenomena are all of one piece and are best described in terms of all accompanying events and circumstances. Thus, although organization requires a book to be compartmentalized, we want to help correlate and integrate what its separate sections can do so that our students will appreciate the unity and interrelatedness of chemistry as the human endeavor it truly is.

SUPPLEMENTARY MATERIALS

From among the rapidly expanding lists of films, charts, transparencies, and other kinds of instructional aids, a few representative items are cited for each unit. Selected references should help the faster moving students pursue independent projects, do collateral reading, or otherwise advance themselves in the pursuit of chemistry. Some of the references may enable teachers to bolster or extend their comprehension and appreciation of the ideas featured in the chapter

ACKNOWLEDGMENTS

It is always difficult when writing a book of this type to give proper recognition to all of the men and women whose efforts to advance the teaching of chemistry make the book possible. In so many ways, chemistry teaching as we know it today is the yield from countless instances of ideas conceived and explored by the total, world-wide fraternity of chemists and chemistry teachers.

Going through our profession's literature that has accumulated year by year, however, reveals the names of a few men who have been especially involved in the creation and adaptation of instructional activities particularly suited to teaching high school chemistry. Hubert N. Alyea of Princeton University is a good example. So is Frederic B. Dutton of Michigan State University. These men have made stellar contributions to the teaching of chemistry through their "Tested Demonstrations in General Chemistry," "Demonstration Abstracts," and "Chem Ed Tested Demonstrations." Less widely publicized, perhaps, but highly valuable are Denman Evan's "Quickies—A Collection of Classroom Demonstrations and Devices for the Teaching of Science."

There is also the urge to acknowledge the work done by the editors of magazines and journals devoted to the teaching of chemistry. The many, many references to the *Journal of Chemical Education,* for example, or the newer *Chemistry,* or the more general, *The Science Teacher,* prompt us to say thanks to such men as William F. Kieffer, Ted Benfey, and Robert Carleton and their assistants.

Finally, we must acknowledge the help we have received from the men and women in recent years who tackled the job of updating and refurbishing high school chemistry courses. Although passing years may dim the meaning of such symbols as CHEMS and CBA, there is little doubt that these projects will continue to influence the instructional activities whereby young people are introduced to the pursuit of chemistry. They have certainly influenced the teaching of the authors of this book, hence, the selection of activities we have included.

We would be remiss if we did not add a note of thanks to those high school boys and girls who have served as laboratory assistants not only while this book was being written but throughout all the years we have been trying out new laboratory exercises, demonstrations, and other classroom activities. They have let us know what will and won't work, what is and what isn't effective. More importantly, their contagious enthusiasm for trying the untried and exploring the yet to be explored has been our principal source of inspiration.

John H. Woodburn
Lewis R. Sanford
W. James Brawley, Jr.

CONTENTS

1. Classifying Kinds of Matter 17

INSTRUCTIONAL ACTIVITIES: STUDENT 17

1-1. GETTING ACQUAINTED WITH THE EARTH'S ELEMENTS 17

1-2. RELATIVE ABUNDANCE OF ELEMENTS, COMPOUNDS, AND MIXTURES 18

1-3. MIXTURES VERSUS COMPOUNDS 18

1-4. RECOGNIZING DIFFERENCES IN PROPERTIES 19

1-5. WHAT HAPPENS TO THE PROPERTIES OF AN ELEMENT WHEN IT BECOMES A PART OF A COMPOUND? 19

INSTRUCTIONAL ACTIVITIES: TEACHER 20

1-T1. ELEMENTS BEFORE AND AFTER FORMING COMPOUNDS OR MIXTURES 21

RELATING OTHER CONCEPTS 23

AUDIO-VISUAL AIDS 23

IDEAS FOR ADDITIONAL INSTRUCTIONAL ACTIVITIES 24

2. Determining Quantity of Matter 25

INSTRUCTIONAL ACTIVITIES: STUDENT 26

2-1. BALANCE SENSITIVITY 26

2-2. BALANCE SENSITIVITY USING CENTIGRAM BALANCE 26

2-3. COMPARING THE PRECISION OF WEIGHT AND VOLUME MEASUREMENTS 26

2-4. USING MASS AND VOLUME MEASUREMENTS TO DETERMINE DENSITY 27

INSTRUCTIONAL ACTIVITIES: TEACHER 27

2. Determining Quantity of Matter *(Continued)*
2-T1. visualizing relationships between mass, volume and density 27
2-T2. seeing through a supposedly puzzling situation 28
2-T3. the specific gravity of "difficult" materials 28
2-T4. using specific gravity to predict solubility 28
2-T5. the density of air 29
audio-visual aids 29
ideas for additional instructional activities 30

3. The Changing Properties of Substances Which Accompany Chemical Changes 33
instructional activities: student 34
3-1. recognizing the evidence of chemical changes 34
instructional activities: teacher 36
3-T1. the disappearing precipitate 36
audio-visual aids 37
ideas for additional instructional activities 37

4. The Particle Nature of Matter 41
instructional activities: student 42
4-1. the idea of relative versus actual weights 42
4-2. using relative weights to count molecules 43
4-3. realizing how large a number the mole is 44
instructional activities: teacher 45
4-T1. man's quest for an effective mental model of the atom 45
4-T2. interpreting observations from the perspective of the particle nature of matter 46
4-T3. visualizing the mole as a quantitative unit 46
audio-visual aids 47
ideas for additional instructional activities 48

5. Communication in Chemistry 49

INSTRUCTIONAL ACTIVITIES: STUDENT 50

5-1. SYMBOLS TO REPRESENT ATOMS 50

5-2. FORMULAS TO REPRESENT SUBSTANCES OR, MORE PRECISELY, PARTICLES OF SUBSTANCES 50

5-3. TRANSLATING CHEMICAL EQUATIONS 51

5-4. MAKING A GAME OF LEARNING SYMBOLS AND FORMULAS 51

INSTRUCTIONAL ACTIVITIES: TEACHER 52

5-T1. CHEMICAL EQUATIONS IN STYROFOAM 52

AUDIO-VISUAL AIDS 53

IDEAS FOR ADDITIONAL INSTRUCTIONAL ACTIVITIES 53

6. Stoichiometry 55

INSTRUCTIONAL ACTIVITIES: STUDENT 55

6-1. EQUATIONS AS "RECIPES" FOR CHEMICAL REACTIONS 55

6-2. VOLUMES AS WELL AS WEIGHTS FROM CHEMICAL EQUATIONS 57

6-3. DETERMINING THE "RECIPE" WHEN YOU CAN ANALYZE THE PRODUCT 57

6-4. COMPARING ACTUAL WITH THEORETICAL YIELDS 58

6-5. COMPARING ACTUAL WITH THEORETICAL YIELDS WHEN THE REACTANTS ARE IN SOLUTIONS 60

6-6. USING STOICHIOMETRY TO FIND THE CORRECT "RECIPE" 61

6-7. USING STOICHIOMETRY TO SOLVE A CHEMIST'S PROBLEM 61

INSTRUCTIONAL ACTIVITIES: TEACHER 62

6-T1. DISCOVERING THE VALUE OF THE COEFFICIENTS IN A CHEMICAL EQUATION 62

6-T2. TAKING STUDENTS THROUGH A "WEIGHT-VOLUME" PROBLEM 63

AUDIO-VISUAL AIDS 63

IDEAS FOR ADDITIONAL INSTRUCTIONAL ACTIVITIES 64

7. Properties of Gases 66

INSTRUCTIONAL ACTIVITIES: STUDENT 67

7. PROPERTIES OF GASES *(Continued)*
7-1. RELATIONSHIP BETWEEN VOLUME AND PRESSURE OF A TRAPPED QUANTITY OF GAS 67
7-2. RELATIONSHIP BETWEEN VOLUME AND TEMPERATURE OF A TRAPPED QUANTITY OF GAS 68
7-3. THE VOLUME OCCUPIED BY A MOLE OF GAS PARTICLES 68
INSTRUCTIONAL ACTIVITIES: TEACHER 69
7-T1. THE DISPLACEMENT RATES OF GASES 69
7-T2. THE PROPERTIES OF GASES AND THE "THIRSTY" BIRD 70
7-T3. THE VAPOR PRESSURE OF LIQUIDS 70
AUDIO-VISUAL AIDS 71
IDEAS FOR ADDITIONAL INSTRUCTIONAL ACTIVITIES 72

8. PERIODICITY IN THE PROPERTIES OF THE ELEMENTS 74

INSTRUCTIONAL ACTIVITIES: STUDENT 74
8-1. CLASSIFICATION IN CHEMISTRY VERSUS IN BIOLOGY 74
8-2. PERIODICITY IN THE PROPERTIES OF THE ELEMENTS AS AN EVOLVING IDEA 75
8-3. RECOGNIZING PERIODICITY IN THE PROPERTIES OF THE ELEMENTS 75
8-4. TRENDS IN PROPERTIES SHARED BY A FAMILY OF ELEMENTS 76
INSTRUCTIONAL ACTIVITIES: TEACHER 77
8-T1. PERIODICITY IN THE PROPERTIES OF THE ELEMENTS SHOWN WITH STYROFOAM MODELS 77
8-T2. TRENDS IN THE PROPERTIES OF THE MEMBERS OF THE ALKALI METAL FAMILY OF ELEMENTS 77
8-T3. TRENDS IN THE PROPERTIES OF THE MEMBERS OF THE HALOGEN FAMILY OF ELEMENTS 78
AUDIO-VISUAL AIDS 81
IDEAS FOR ADDITIONAL INSTRUCTIONAL ACTIVITIES 81

9. SOLUTIONS AND SOLUBILITIES 83

INSTRUCTIONAL ACTIVITIES: STUDENT 84
9-1. HANDBOOK INFORMATION DEALING WITH SOLUTIONS AND SOLUBILITIES 84

9. SOLUTIONS AND SOLUBILITIES *(Continued)*

9-2. ELECTRICAL CONDUCTIVITY AS A CLUE TO THE NATURE OF SOLUTIONS AND SOLUBILITIES 85

9-3. SOLUTIONS OF SUBSTANCES VERSUS SUBSTANCES IN THE LIQUID STATE 86

9-4. SOLUTION: AN ENERGY INVOLVING PROCESS 87

9-5. INTERACTIONS BETWEEN SOLUTIONS 88

9-6. ADDITIONAL INTERACTIONS BETWEEN SOLUTIONS 89

INSTRUCTIONAL ACTIVITIES: TEACHER 90

9-T1. CALCULATIONS INVOLVING SOLUBILITY PRODUCTS 90

9-T2. THE COLLIGATIVE PROPERTIES OF SOLUTIONS 91

9-T3. THE SOLUTION OF HYDROGEN CHLORIDE IN WATER 92

AUDIO-VISUAL AIDS 92

IDEAS FOR ADDITIONAL INSTRUCTIONAL ACTIVITIES 95

10. ACIDS AND BASES 98

INSTRUCTIONAL ACTIVITIES: STUDENT 99

10-1. "ANYTHING WATER CAN DO, ACIDS AND/OR BASES CAN DO BETTER" 99

10-2. GETTING ACQUAINTED WITH ACIDS 99

10-3. GETTING ACQUAINTED WITH BASES 100

10-4. THE INTERACTION OF ACIDS AND BASES 100

10-5. PRECISE DETERMINATION OF ACID OR BASE CONCENTRATIONS 101

INSTRUCTIONAL ACTIVITIES: TEACHER 102

10-T1. INDICATORS AND THE STRENGTH OF ACIDS AND BASES 103

10-T2. THE BRONSTED-LOWRY CONCEPT OF ACID-BASE INTERACTION 104

10-T3. ACID-BASE INTERACTIONS AND THE BUFFERING PHENOMENON 105

10-T4. A COMPUTERIZED LESSON INVOLVING TITRATION CURVES 106

10-T5. COMPUTER-ASSISTED INSTRUCTION INVOLVING SIMULATED TITRATION 112

AUDIO-VISUAL AIDS 114

IDEAS FOR ADDITIONAL INSTRUCTIONAL ACTIVITIES 115

11. Oxidation—Reduction 118

INSTRUCTIONAL ACTIVITIES: STUDENT 119
11-1. COMPETITION FOR ELECTRONS 119
11-2. COMPETITION FOR ELECTRONS BY THE MEMBERS OF THE HALOGEN FAMILY 120
INSTRUCTIONAL ACTIVITIES: TEACHER 120
11-T1. STEP-BY-STEP OXIDATION AND REDUCTION 120
AUDIO-VISUAL AIDS 123
IDEAS FOR ADDITIONAL INSTRUCTIONAL ACTIVITIES 123

12. Electrochemistry 125

INSTRUCTIONAL ACTIVITIES: STUDENT 125
12-1. QUANTITATIVE DESCRIPTION OF ELECTRON HOLDING OR ELECTRON TAKING PROPERTIES OF THE ELEMENTS 125
12-2. PREDICTING THE VOLTAGE OF VARIOUS COMBINATIONS OF HALF CELL REACTIONS 127
12-3. USING "COMMERCIAL" ELECTROCHEMICAL CELLS TO MANAGE OR CONTROL CHEMICAL REACTIONS 128
INSTRUCTIONAL ACTIVITIES: TEACHER 129
12-T1. INTERACTIONS BETWEEN ELECTRICAL AND CHEMICAL PHENOMENA 129
12-T2. FARADAY'S ELECTROCHEMICAL LAWS 130
AUDIO-VISUAL AIDS 131
IDEAS FOR ADDITIONAL INSTRUCTIONAL ACTIVITIES 132

13. Bonds Within Molecules 134

INSTRUCTIONAL ACTIVITIES: STUDENT 135
13-1. A LESSON IN WORDS AND WHAT THEY CAN AND CAN'T DO FOR US 135
INSTRUCTIONAL ACTIVITIES: TEACHER 136
13-T1. OBSERVATIONS VERSUS DEDUCTIONS VERSUS MENTAL MODELS OR THEORIES 136
AUDIO-VISUAL AIDS 138
IDEAS FOR ADDITIONAL INSTRUCTIONAL ACTIVITIES 138

14. Bonds Between Molecules 140

INSTRUCTIONAL ACTIVITIES: STUDENT 141

14. BONDS BETWEEN MOLECULES *(Continued)*
14-1. RELATING THE "STICKINESS" BETWEEN MOLECULES TO OTHER PROPERTIES OF SUBSTANCES 141
INSTRUCTIONAL ACTIVITIES: TEACHER 143
14-T1. CONFRONTING STUDENTS WITH OBSERVATIONS INVOLVING SELECTED BOND FORMING, STRETCHING, OR BREAKING PHENOMENA 143
AUDIO-VISUAL AIDS 144
IDEAS FOR ADDITIONAL INSTRUCTIONAL ACTIVITIES 144

15. METALS AND THE METALLIC BOND 146

INSTRUCTIONAL ACTIVITIES: STUDENT 146
15-1. METALS VERSUS NONMETALS 146
15-2. WINNING METALS FROM THEIR ORES 147
15-3. THE CORROSION OF METALS 148
AUDIO-VISUAL AIDS 150
IDEAS FOR ADDITIONAL INSTRUCTIONAL ACTIVITIES 151

16. QUANTUM THEORY 153

INSTRUCTIONAL ACTIVITIES: STUDENT 154
16-1. ENERGY, TOO, COMES IN PARTICLES 154
16-2. THE "ADDRESS" OF EACH ELEMENT'S ELECTRONS 155
INSTRUCTIONAL ACTIVITIES: TEACHER 156
16-T1. FORMULAS AND EQUATIONS 156
IDEAS FOR ADDITIONAL INSTRUCTIONAL ACTIVITIES 158

17. REACTION MECHANISMS AND THE KINETICS OF REACTIONS . 159

INSTRUCTIONAL ACTIVITIES: STUDENT 160
17-1. SOLIDS VERSUS SOLUTIONS 160
17-2. GUESSING OR HYPOTHESIZING THE MECHANISM OF A REACTION 160
INSTRUCTIONAL ACTIVITIES: TEACHER 161
17-T1. THE TRADITIONAL IODINE STARCH CLOCK REACTION 161
17-T2. FACTORS AFFECTING REACTION RATES 162
17-T3. AUTOCATALYSIS 164
17-T4. A SLIGHTLY "SNEAKY" MECHANISM 164
AUDIO-VISUAL AIDS 164
IDEAS FOR ADDITIONAL INSTRUCTIONAL ACTIVITIES 165

18. Equilibria Systems 167

INSTRUCTIONAL ACTIVITIES: STUDENT 168
18-1. COMPARING SETS OF CHEMICAL SYSTEMS 168
18-2. PUTTING EQUILIBRIA SYSTEMS UNDER STRESS 169
18-3. THE COMMON ION EFFECT 170
INSTRUCTIONAL ACTIVITIES: TEACHER 170
18-T1. STRESSES APPLIED TO GASEOUS SYSTEMS AT EQUILIBRIUM 170
AUDIO-VISUAL AIDS 171
IDEAS FOR ADDITIONAL INSTRUCTIONAL ACTIVITIES 171

19. Thermochemical Phenomena 173

INSTRUCTIONAL ACTIVITIES: STUDENT 174
19-1. TEMPERATURE VERSUS HEAT CONTENT 174
19-2. TEMPERATURE CHANGES WITH A DIFFERENCE 174
19-3. PROPERTIES OF METALS BEFORE AND AFTER BEING HEATED 175
19-4. HEAT TO START VERSUS HEAT TO CONTINUE A REACTION 175
19-5. THE HEAT ADDITIVE OR HESS'S LAW 176
INSTRUCTIONAL ACTIVITIES: TEACHER 177
19-T1. INTERACTIONS OF ELECTRODE POTENTIAL DIFFERENCE, DELTA-H, DELTA-G, DELTA-S, AND ENTROPY 177
AUDIO-VISUAL AIDS 179
IDEAS FOR ADDITIONAL INSTRUCTIONAL ACTIVITIES 179

20. Carbon Chemistry 181

INSTRUCTIONAL ACTIVITIES: STUDENT 182
20-1. STRUCTURAL FORMULAS IN STYROFOAM 182
20-2. ORGANIC COMPOUNDS FROM THE OXIDATION SERIES POINT OF VIEW 182
20-3. MANAGING OR MANIPULATING ORGANIC REACTIONS 183
20-4. DEDUCING THE STRUCTURE OF ORGANIC MOLECULES 184
INSTRUCTIONAL ACTIVITIES: TEACHER 185
20-T1. CIS-TRANS ISOMERISM 185

20. CARBON CHEMISTRY *(Continued)*
AUDIO-VISUAL AIDS 186
IDEAS FOR ADDITIONAL INSTRUCTIONAL ACTIVITIES 187

21. BIOCHEMISTRY 189

INSTRUCTIONAL ACTIVITIES: STUDENT 190
21-1. BIOCHEMISTRY IN STYROFOAM 190
INSTRUCTIONAL ACTIVITIES: TEACHER 191
21-T1. BIOCHEMISTRY IN CURRENT LITERATURE 191
AUDIO-VISUAL AIDS 193
IDEAS FOR ADDITIONAL INSTRUCTIONAL ACTIVITIES 193

22. DESCRIPTIVE CHEMISTRY 195

INSTRUCTIONAL ACTIVITIES: STUDENT 196
22-1. SOME THOUGHTS ABOUT THE ELEMENT, CARBON 196
22-2. THE PROPERTY OF SOLUBILITY 197
INSTRUCTIONAL ACTIVITIES: TEACHER 198
AUDIO-VISUAL AIDS 200
IDEAS FOR ADDITIONAL INSTRUCTIONAL ACTIVITIES 201

23. ANALYSIS AND SYNTHESIS 202

INSTRUCTIONAL ACTIVITIES: STUDENT 203
23-1. QUALITATIVE ANALYSIS 203
23-2. SYNTHESIS OF SERPENT'S EGGS 204
AUDIO-VISUAL AIDS 206
IDEAS FOR ADDITIONAL INSTRUCTIONAL ACTIVITIES 207

24. NUCLEAR CHEMISTRY 209

INSTRUCTIONAL ACTIVITIES: TEACHER 210
24-T1. GOVERNMENTAL SUBSIDY OF NUCLEAR VERSUS FOSSIL FUEL RESOURCES—A STUDENT DEBATE 210
AUDIO-VISUAL AIDS 210
IDEAS FOR ADDITIONAL INSTRUCTIONAL ACTIVITIES 211

25. GEOCHEMISTRY 213

INSTRUCTIONAL ACTIVITIES: STUDENT 214
25-1. PURE SALT FROM A PREHISTORIC OCEAN 214
IDEAS FOR ADDITIONAL INSTRUCTIONAL ACTIVITIES 215

Appendix A

Addresses for obtaining audio-visual aids 217

Bibliography 219

Index to Authors 221

General Index 226

1

CLASSIFYING KINDS OF MATTER

Our students come to us accustomed more to classifying substances on the basis of their uses than on similarities in their properties. We move abruptly to a classification system which features the properties of substances. We expect our students to appreciate this system increasingly as they come to see that these properties are related to the kinds, numbers, and arrangements of the fundamental particles of which the substances are composed.

As the course evolves, we hope our students will grasp the elegance and simplicity inherent in reducing all of the world's myriad substances to being composed of no more than 100 or so different kinds of particles—the atoms of the elements. Continued references to the particles of substances leads to clear-cut meanings for such terms as atoms and molecules, elements, compounds, and mixtures.

Always a problem in chemistry, the names of substances should take on additional meaning. Names derived from the source, preparation, or special properties of substances must be taken for what they are. A larger goal is the recognition of attempts to introduce logical and orderly systems for naming substances.

INSTRUCTIONAL ACTIVITIES: STUDENT

1-1. Getting Acquainted with the Earth's Elements. Select the 10 elements whose names begin with the letter closest alphabetically to the first letter of your last name. Consult the *Handbook of Chemis-*

try and Physics or a good encyclopedia and determine which of these elements are to be found in nature predominantly in the free state, that is, uncombined with other elements in the form of compounds. Also determine whether these elements exist as gases, liquids, or solids under usual environmental temperatures and pressures. Be ready when you present this information to the class to discuss any relationships which you think might exist between where the element can be found; the state in which it exists, that is, solid, liquid, or gas; and its chemical properties.

1-2. Relative Abundance of Elements, Compounds, and Mixtures. Remind yourself that elements consist, basically, of only one kind of atom; compounds consist of molecules in which two or more kinds of atoms are chemically combined in definite ratios; and mixtures consist of two or more elements or compounds combined in no definite ratio. Survey the shelves of the chemistry storeroom, a drugstore, hardware, supermarket, or department store and estimate what fraction of the items on the shelves fall into each category—element, compound, or mixture.

Think about the implications in the differences in the fractions in each of the different situations. For example, why might you find more items in the element category in hardwares than in supermarkets? Or why should there be more compounds on chemistry storeroom shelves than on drugstore shelves?

1-3. Mixtures Versus Compounds. Obtain a 6 to 8 gram sample of a plant fertilizer. Dry the sample as best you can and reweigh. Add 25 ml of water to the sample and stir for 5 minutes or so. Filter, dry the insoluble portion on the filter paper, weigh and calculate the per cent this weight is of the original dried sample. Evaporate the filtrate, dry, weigh, and calculate the per cent this weight is of the weight of the original dried sample of the fertilizer.

Prepare a mixture consisting of 2 grams each of baking soda, tartaric acid, and cream of tartar. Place in a 250 ml beaker and add 25 ml of water. Stir gently for 5 minutes or so. As you did with the sample of fertilizer, filter, dry the precipitate if any appears, and evaporate the water from the filtrate. Calculate the per cent the weights of the filtrate after evaporation and the dried precipitate are of the original weight of the soda, tartaric acid, and cream of tartar mixture.

Do the data from the fertilizer differ markedly from the data from the second mixture? Could you restore the original mixtures by simply recombining the dried precipitates and evaporated filtrates? Did you see anything happening when you added the water to the second mixture that might explain why the data from this mixture differed from the data from the fertilizer sample?

1-4. Recognizing Differences in Properties. Weigh a new galvanized (zinc coated) iron nail and then put it in about 20 ml of 1 M hydrochloric acid. Observe at times during the lab period. On the following day, observe the nail in the acid then remove, dry, and weigh. On the basis of your observations including change in weight, which of the terms—element, compound, or mixture—best describes the original nail? Do you feel you need an additional term?

Starting with another new galvanized nail, weigh and place in about 10 ml of 6 M nitric acid. Do this under the *fume hood.* Observe at times and leave until the following day. Assuming that by then the nail has disappeared completely, transfer the residue to an evaporating dish using a wash bottle to rinse the beaker into the evaporating dish. After the liquid portion of the residue has evaporated completely, examine the crystalline residue closely. Assuming you find two kinds of crystals in the residue, describe as accurately as you can how the two acids, hydrochloric and nitric, react with the nail.

1-5. What Happens to the Properties of an Element When it Becomes a Part of a Compound? Examine a few crystals of silver nitrate (*poison and corrosive*). Dissolve the crystals in a few milliliters of water and observe the solution. If you forgot to use distilled water, do not discard your solution but go ahead and compare your results with this exercise with a neighbor who did use distilled water. Polish the end of a copper wire, place your solution of silver nitrate in a test tube, and let the copper wire stand in the solution. Observe during the remainder of the lab period.

On the following day, remove the copper wire as gently as possible and place on a watch glass. Examine the wire, the attached material, and the original silver nitrate solution. Without reading ahead, try to explain everything that happened during the reactions. Supposedly, some of the copper from the wire went into solution and the silver from the silver nitrate came out of solution in the form of tiny crystals of metallic silver.

Pour the resulting blue solution of copper and, perhaps, silver nitrate into a fresh test tube and drop into the test tube a shiny iron nail. Again, observe during the lab period and leave over night. On the following day, observe the results and compare with those you found after leaving the copper wire in the original silver nitrate solution.

In general, does an element retain its properties when it becomes a component of a compound? Can an element regain its properties after it has lost its properties within a compound? Is it really accurate to say that an element can "lose" or "regain" its properties? Would it be better to say that among the properties of an element are the various ways it may affect the properties of its compounds? And here is a question that you won't be able to answer until much, much later in your pursuit of chemistry: How does a compound get the properties which, sometimes, are not even hinted at by the properties of the separate elements which produce the compound?

INSTRUCTIONAL ACTIVITIES: TEACHER

The instructional activities that are used to clarify the ideas of elements, compounds, and mixtures usually come early in the course. Because of the importance of first impressions, these activities take on added importance. Particularly sensitive are the attitudes our students develop regarding the role the laboratory is to play in their study of chemistry. They begin deciding quite early whether the lab is to be a place to work or a place to loaf.

A vital but more subtle aspect of student attitude toward laboratory instruction involves the "discovery" versus the "reinforcement" feature of laboratory exercises. Some teachers work toward having their students discover orderly relationships, principles, and concepts by analyzing firsthand their own observations of selected chemical reactions. Other teachers prefer presenting carefully thought through interpretations of principles and concepts and then using the laboratory to clarify or enhance their understanding and appreciation.

Seldom do instructional activities need to be exclusively "discovery" or "reinforcement" oriented. Many teachers begin with the more comfortable "reinforcement" approach. As rapidly as their students can gain the necessary capabilities, the skills and attitudes, needed to enable the "discovery" oriented type of laboratory work to be administered effectively, teachers can move their students increasingly in this direction. Obviously, to do so prematurely will very likely lead to prompt abandonment of the "discovery" approach.

1-T1. Elements Before and After Forming Compounds or Mixtures. Provide each student with random weights of sulfur and iron filings. This can be done efficiently by scooping appropriate amounts onto one quarter 8½ × 11 inch pieces of paper on the check-out table. The uniform weight of these paper "carriers" provides a quick check on students who are not using their balances correctly or the balance that is out of order. And the papers are easily folded to provide pouring funnels.

Advise your students to record the weights of each element and to then mix them thoroughly in a previously weighed crucible with cover. Have them heat the crucibles (*hood*) moving the cover on and off with crucible tongs so as to observe the reaction without unduly risking the loss of the contents during the heating process. When no further action seems to be taking place, remove the covers and continue heating until cover and crucible are clean except for the residue in the bottom of the crucible. As soon as the crucibles are cool and weighed, transfer a few crumbs of the residue to a small test tube labelled with name and/or lab station number and deliver to a designated container at the check-out table. The remainder of the residue is to be retained in the crucibles.

Prepare a data table on the chalkboard or overhead projector to display such student data as: (1) initial weight of iron, (2) initial weight of sulfur, (3) weight of residue, and (4) weight of sulfur retained in the residue, assuming no iron was lost during the reaction. Plan this data table so that the following columns can be added during the next day's class session: (5) reaction of the residue to a magnet, (6) evidence of recrystallized sulfur following addition of carbon disulfide to residue and evaporation, and (7) evidence of blackening of lead acetate paper after adding hydrochloric acid to the original residue.

Items 1 through 4 of this data table can be entered by the students during the first day of this exercise. To obtain the information for items 5, 6, and 7, divide the day's collection of test tubes into two groups. To one group, add 1 ml or so of carbon disulfide and place where the solvent can evaporate away safely before the next class meeting. To the second group, add 1 ml or so of 1 M hydrochloric acid, hang a moistened strip of lead acetate paper over the mouth of the test tube, and stopper loosely with an absorbent cotton plug. If all of these operations are done under the hood, student exposure to hydrogen sulfide can be eliminated or, at least, kept to an absolute minimum.

On the second day to be devoted to this exercise, have your students enter their individual observations for columns 5 and 6 or 7. If the magnet is kept in a sealed plastic bag, its poles will not become "contaminated" with iron filings from previous use.

Have your students prepare a data table that summarizes the group's data, a step that can be aided by using a dittoed form. This form can also look ahead to the "analysis of data" phase of the exercise by listing an array of hypotheses against which each student is to project his own as well as the consensus of the group's data. Sample hypotheses are: (1) Elements may lose their physical properties when forming compounds. (2) Elements may lose their chemical properties when forming compounds. (3) Compounds may have properties other than those of the individual elements which make up the compound. (4) Any weight of one element will combine to form a compound with any weight of another element with which it reacts. (5) Energy is either taken in or given off whenever elements combine to form compounds. (6) Any action that will separate a mixture of elements will also separate these same elements after they have combined to form a compound.

The role of the teacher during the "analysis of data" phase of this kind of approach to laboratory exercises depends pretty much on the attitudes and habits your students have acquired from previous science laboratory activities. Some students may have come to look upon lab work as being nothing more than an opportunity to see things happen with no tendency to engage their minds in any kind of disciplined thinking processes or procedures. A pronounced resistance to any notetaking is also a part of this syndrome. To these students, the lab exercise is all wrapped up as soon as the flames die down and the smoke clears away.

If the teacher senses that his students are coming to him without effective attitudes and work habits, he should exploit the favorable "teachability" of these students which may prevail during the getting acquainted period when everyone seems to be on at least temporarily good behavior. This is why we feature highly structured record keeping and analysis of data activities in the design of this introductory lab exercise. It is up to the finesse of the teacher to have his students realize that all of the followup work associated with the exercise is essential to the ultimate appreciation of what the exercise was all about and what it was intended to add to their evolving mastery of the chemistry course.

Obviously, finesse cannot be prescribed. Sometimes it is a matter of being able to ask the right kinds of leading questions. For example, some students respond nicely to such a question as: "I wonder if all residues contained the same percentage of iron and sulfur, again assuming that no iron escaped during the reaction? If not, do you suppose there is something here that is related to the original weight ratios between the two elements? Is it easier to find an orderly relationship in our data if we include only the data from students whose residues neither responded to a magnet nor showed evidence of retaining uncombined sulfur but did discolor lead acetate paper?"

RELATING OTHER CONCEPTS

The ideas involved in this chapter will be retrieved and revisited quite often as the material evolves. There may be less occasion to expand or enrich their treatment by reaching ahead for additional material. It may be worthwhile, however, to consider the following student and/or teacher instructional activities:

Audio-Visual Aids

1-AV1. "Chemistry for Today." Twelve filmstrips with approximately 50 frames per strip. Society for Visual Education. Introductory survey.

1-AV2. "Chemistry: Matter and Chemical Change." Programmed Instruction. Approximately 3000 frames in reusable book. Teaching Materials Corporation. Interpretations of valence, bonding, periodicity, atomic structure, etc.

1-AV3. "Elements, Compounds, Mixtures." Color. 30 min. Coronet. Features the iron-sulfur reaction.

1-AV4. "High School Chemistry." Videotape. 94 lessons. National Instructional Television. Planned to supplement traditional instruction, each lesson is devoted to one of the fundamental principles of chemistry.

1-AV5. "Preface to Chemistry." Color. 16 min. Britannica. Introduction to chemistry.

Ideas for Additional Instructional Activities

1-AD1. Dinga, Gustav P., "The Elements and the Derivation of Their Names and Symbols." *Chemistry* 40: 20–21. February, 1968. Tabular presentation.

1-AD2. In activity 1-T1, copper, lead, nickle, or cadmium may be used in lieu of iron. Copper foil works well. If a sheet is weighed and then cut into square centimeter pieces, different students can be given various multiples of approximately the same weight. This approach enables the laboratory instructor to spot the students who are making obvious errors in reading their balances. In contrast to relying on magnetic properties to show when the iron has been converted to the sulfide, when other metals are used, the ease with which the sulfide product can be crumbled indicates that a reaction has occurred.

1-AD3. Learning the names and combining capacities of the atoms and polyatomic ions commonly encountered in high school chemistry can be aided by an easily constructed card game. Print the symbol or formula of each ion on each of four cards. Omit the sign and magnitude of charge. Although students may make up their own rules, one type of game involves dealing each player five cards. The remaining cards are placed face down with one card turned face up. The first player may take the exposed card or draw from the unexposed deck. The object of the game is to make compounds with the cards in each player's hand. When a compound is laid down, it must be named. If the formula is not correct or if it is misnamed, the corrector scores points and the faulting player must return the cards to his hand. The game ends when a player is out of cards. Scoring is optional but may be based on total number of compounds made, total number of cards used in forming compounds, and so forth.

2

DETERMINING QUANTITY OF MATTER

To recognize or discover almost any kind of orderliness in their observations of a chemical event, our students must be able to use measuring devices accurately and intelligently, especially balances. This is especially true when their measurements of macroscopic properties are to be used to deduce what is happening at the atom or molecule level. Furthermore, they must acquire the ability to determine whether or not an orderly relationship that is actually present in a set of observations has been compromised by inadequate accuracy or precision in their measuring instruments. Overly impressed, perhaps, by the sensitiveness of chemical balances or the apparent precision of the markings on graduated cylinders, students need to be reminded that their most careful weighings or volume determinations may need to be interpreted as within plus or minus as many as 10^{18} atoms or molecules.

A comment on the use of *weight* versus *mass* or *weighing* versus *massing* in school chemistry courses. Obviously, students who are to continue their education in the sciences must make the necessary transition in the use of these terms. For general education students, however, the use of the terms, *mass* and *massing,* seems to widen the gap between the classroom and everyday appreciation of chemistry. For this reason, many teachers prefer to clarify the situation but then continue to use the less precise terminology.

INSTRUCTIONAL ACTIVITIES: STUDENT

2-1. Balance Sensitivity. Without counting them as you do so, use a platform balance sensitive to 0.1 gram to weigh 1 gram of the small objects (sequins) at your lab station. Now dump them on the table top and count them by sorting them into piles of 10. Repeat the process of weighing then counting 1 g samples of sequins 9 more times. Visit 9 of your neighbors and record the number of sequins they found in their first weighing of a 1 g sample. Include your own first weighing and calculate the average number of sequins in 1 gram as determined by pooling the efforts of 10 students. Calculate the average of your own 10 weighings of 1 g samples of the sequins. How does the average of 10 of your own weighings compare with the average of 10 students' single weighings?

Repeat these activities using the smaller objects (glitter stars) but this time weigh and count only 0.3 g samples.

Has your balance been weighing plus or minus 0.1 g? To answer this question, calculate the number of sequins needed to weigh 0.1 gram by using the average number in 1.0 gram samples and check to see how many of your individual weighings of 1.0 g samples yielded numbers of sequins within plus or minus this number.

Suppose each of your classmates carefully counted out 100 sequins and mixed them with 50 equally carefully counted glitter stars. If you were to separate but not count the sequins and the glitter stars, could you use your decigram balance to "discover" that the mixture actually contained a two to one mixture of sequins and glitter stars? Could you prove that a student had goofed and didn't really prepare a two to one mixture?

2-2. Balance Sensitivity Using Centigram Balance. Using a centigram balance assumed to be sensitive to 0.01 g, repeat Activity 2-1. How does the more sensitive balance affect the data from this exercise?

2-3. Comparing the Precision of Weight and Volume Measurements. Weigh a 100 ml graduated cylinder. Fill to the 100 ml mark using a medicine dropper to add the final few drops. Reweigh. Does the weight of your 100 ml of water agree with the usual 1 g per ml? Was the bottom of the meniscus exactly at the 100 ml mark? If the balance is rated to be sensitive to 0.1 g, can you obtain 100 ml of water more accurately using the balance than by using the graduated cylinder? Consult the *Handbook of Chemistry and Physics* to determine how large an error is introduced when we assume that 1 g of water occupies 1 ml.

Using a 100 ml graduate, fill it to the mark 5 times and pour the water into a 500 ml volumetric flask. How many drops of water must be added or removed from the flask to bring the water level to the 500 ml mark? Again, use the graduate but this time weigh 100 g of water 5 times and pour into the flask. Within how many drops did the 500 grams of water come to filling the flask exactly to the mark? Compare your results with 10 or so of your neighbors and decide what is the most reliable way to use a graduated cylinder.

2-4. Using Mass and Volume Measurements to Determine Density. Obtain 20 to 25 grams of rhombic sulfur in lumps small enough to drop into a 100 ml graduate. Fill a 100 ml graduate to the 50 ml mark and then add the sulfur carefully to avoid splashing. Determine the volume of the sulfur. Remember that a balance should be read as plus or minus so many grams and a graduate should be read as plus or minus so many milliliters. Calculate the density of the sulfur using the maximum negative correction for the balance and the maximum positive correction for the graduate. Recalculate the density but this time use the minimum correction for the balance and the maximum negative correction for the graduate. Is the handbook value for the density of sulfur within the range of your maximum and minimum determinations?

INSTRUCTIONAL ACTIVITIES: TEACHER

It is often a matter of choice whether an instructional activity will be turned over to the students or carried out pretty much by the teacher, a choice that applies to many of the activities described in this book. The teacher may take over an activity when he wants to emphasize something the students might overlook. Sometimes delicate apparatus or fragile glassware is better handled by the teacher. And some activities allow the teacher to add a bit of drama or showmanship to routine class sessions.

2-T1. Visualizing Relationships Between Mass, Volume, and Density. Adjust several 100 ml or larger graduates to have equal mass by taping bits of sheet lead to their bottoms. Show your students that all of these graduates weigh the same. Put a small volume of mercury in one graduate and show the volumes of other liquids needed to achieve equal weights. Thinking ahead to 2-T2, be sure to include water colored brightly with a food dye, kerosene, and trichlorotrifluorethane or some other less toxic but equally dense liquid.

2-T2. Seeing Through a Supposedly Puzzling Situation. Without tipping off your students that this is a follow-up of 2-T1 and with a minimum of "sound track," pour 2 inches or so of mercury into a large graduate. Follow this with equal volumes of chloroform, water, and kerosene. Kerosene and chloroform being miscible, avoid allowing the kerosene to splash through the water layer. Before dropping a cube of lead, a marble, an ebony block, and a cube of cork into the graduate, announce that you are about to confront your students with a kind of puzzling situation and ask them to avoid spoiling the fun and to not blurt out the explanation but to write their individual "hypotheses" so each person's thinking can be checked privately. This format emphasizes the value of retrieving previously taught information. A more abstract version of the same stunt is arranged by camouflaging some of the components. For example, the water can be colored with food dye and the trichlorotrifluorethane with iodine.

2-T3. The Specific Gravity of "Difficult" Materials. Determine the mass of a pycnometer filled with distilled water. Pour out the water and stuff in a piece of nylon cloth. Add water to fill the unfilled space. Replace stopper and weigh. Convince your students that the specific gravity of the nylon is equal to its weight in air divided by the weight of the pycnometer plus water plus the nylon cloth minus the weight of the water filled pycnometer.

2-T4. Using Specific Gravity to Predict Solubility. In advance, prepare 0.5, 1.0, 2.0, 3.0 M and a saturated solution of sodium chloride. Also, preweigh 5 water filled pycnometers. Confront your students with the increasingly concentrated salt solutions and challenge them to determine the concentration of a saturated salt solution.

Experienced teachers rely on interpreting looks on student faces to determine what level of clues and hints to offer after confronting their students with a challenging assignment. One problem is to be led into diminishing the challenge by students who don't want to think. There is the well known phenomenon of the self-appointed bellwether student whose role is to protect his classmates from having to retrieve knowledge, especially mathematical abilities, by disclaiming, "We haven't been taught how to do this." Sometimes it works to shush the disclaiming student and simply walk around the room and reward the students who are making sincere efforts to meet the challenge by offering private tips and cues all of which, of course, will be overheard.

2-T5. The Density of Air. This activity draws most of its significance from the fact that observations of certain chemical events are likely to be compromised by the "weight of the displaced air."

Evacuate a preweighed 50 ml round bottom, heavy wall flask that is fitted with a one-hole stopper carrying either a glass stopcock or a short length of heavy wall rubber tubing and a hose clamp. Rather than attaching the exhaust pump directly to this flask, you may prefer to use an intermediate 2-liter bottle. A 3-hole stopper in this bottle permits attaching it to the exhaust pump, adding a manometer, and the flask to be evacuated.

Reweigh the evacuated flask. Determine the volume of the flask, or at least that portion of its volume which was "evacuated," by filling a sink or other large container with water and open the hose clamp with the end of the hose under water. Use a graduate or balance to determine the weight of water forced into the flask and, hence, the volume of air represented by the difference between the weight of the flask before and after evacuation.

Challenge your students to scale up their calculated value for the density of air to g/liter. Provide the barometric reading and room temperature, the confidence limits of the measuring instruments used, and determine whether their calculated value agrees or falls into the range of agreement with handbook values.

Audio-Visual Aids

2-AV1. "Air Damped Analytical Balance." Color. 6 min. Yale University. Use and care of.

2-AV2. "Fire Safety in the Laboratory." 3 min. Yale University. Safety measures presented humorously.

2-AV3. "Handling Solids." Color. 3 min. Yale University. Pouring, weighing, disposing, and protecting integrity of containers.

2-AV4. "Introduction to the Chemical Laboratory." Color. 7 min. Chemistry Laboratory Technician. Elementary laboratory techniques.

2-AV5. "Matter and Energy." Color. 11 min. Coronet. Matter in its different forms, physical and chemical change, the laws of conservation of matter and energy.

2-AV6. "Mass and Weight." Color. 11 min. Coronet. The behavior of a bowling ball in a real bowling alley and in imaginary alleys at the North Pole, at the equator, on the moon, and in the depths of intergalactic space.

2-AV7. "The Pipette." Color. 85 frames. Communicable Disease Center. How to avoid hazards in the use of.

2-AV8. "Pipetting, Chemical Techniques." Color. Four titles: Basic Techniques, 5 min; Blow-out Pipettes, 2 min; Pipettors, 5 min; Transfer Pipettes, 3 min. Norwood Films. For training technicians in clinical laboratories.

2-AV9. "Quantitative Transfer." Color. 17 min. Yale University. Using ferric hydroxide for the illustration, shows techniques of filtering, weighing precipitates, etc.

2-AV10. "Safety in the Laboratory." Color. 8 min. Disraeli Films. Pouring acids, heating liquids, cutting glass tubing, smelling vapors.

2-AV11. "Single Pan Analytical Balance." Color. 6 min. Yale University. Use and care of.

2-AV12. "Use of the Triple Beam Balance." 6 min. Purdue University.

2-AV13. "Using a Pipette." Color. 4 min. Yale University. Calibration, cleaning, and precision in use of.

2-AV14. "Using the Laboratory." Color. 11 min. Coronet. Techniques, attitudes, and safety practices.

2-AV15. "Volumetric Measurements." 9 min. Purdue University. Proper use of scale, burette, and graduate.

2-AV16. "Weighing and the Analytical Balance." 15 min. University of Minnesota. Use and care of.

Ideas for Additional Instructional Activities

2-AD1. By substituting liquids other than water in Activity 2-3, especially after the students have established the degree of accuracy with which their graduates can be read, they will appreciate all the more the 1 g/ml situation in using water as a standard.

2-AD2. Borrow rock or mineral specimens from the earth science teacher and adapt Activity 2-4 to show that specific gravity determinations are useful in identifying rock and mineral specimens.

2-AD3. In Activity 2-T5, after establishing the volume of the flask, evacuate as before but allow illuminating gas or some other gas less dense than air to refill the flask. Challenge your students to account for the differences in the weights of the flask before and after being evacuated and after being filled with the gas.

2-AD4. The teacher's manual for the *Chemical Bond Approach* textbook lists the following "Tolerances and Probable Student Accuracy for Use of Volumetric Ware, Balances, and Thermometers."

Burettes, Class B

Capacity, ml	Subdivisions, ml	Manufacturer's tolerances, ml	Probable student reproducibility, ml
25	0.1	±0.06	±0.05
50	0.1	±0.10	±0.05
100	0.2	±0.20	±0.10

Graduated Cylinders, Class B

Capacity, ml	Subdivisions, ml	Manufacturer's tolerances, ml	Probable student reproducibility, ml
10	0.2	±0.1	±0.1
25	0.5	±0.3	±0.3
50	1	±0.4	±0.5
100	1	±0.6	±0.5

Volumetric Flasks, Class B

Capacity, ml	Subdivisions, ml	Manufacturer's tolerances, ml	Probable student reproducibility, ml
25		±0.06	±0.1
50		±0.10	±0.2
100		±0.16	±0.2
250		±0.24	±0.3
500		±0.30	±0.5
1000		±0.60	±1.

Balances

In general, the sensitivity of a balance follows the manufacturer's specifications. Probable student ability to achieve reproducible weighings may be estimated to be plus and minus twice the magnitude of the sensitivity.

Thermometers

—10 to 150 degrees 1° ±1° ±0.5° With Hand Lens: ±0.3°

—10 to 50 degrees 0.1° ±0.5° ±0.05° With Hand Lens: ±0.03°

2-AD5. Astin, Allen V., "Standards of Measurement." *Scientific American,* 218: 50–62. June, 1968. Definitions of units as derived from natural constants.

2-AD6. Neidig, H. A., *et al,* "Statistical Treatment of Experimental Data." Chemistry, Part I, 40: 28–32; Part II, 40: 28–31. November and December, 1967. Using significance tests and confidence limits to distinguish meaningful from meaningless data.

2-AD7. Schubert, Leo, "Units and Dimensional Analysis." *The Science Teacher,* 29: 6–9. March, 1962. Emphasizes the necessity of carrying all units through computations.

2-AD8. Youden, W. J., *Experimentation and Measurement.* New York: Scholastic, 1962. An introduction to the "laws of measurement" written expressly for the high school audience. Includes the statistical treatment of data.

3

THE CHANGING PROPERTIES OF SUBSTANCES WHICH ACCOMPANY CHEMICAL CHANGES

Almost everyone is sure to notice flames and smoke, bubbles or dense precipitates when these flashy observations accompany the rearrangements of atoms during a chemical change. An important measure of our students' progress is their ability to notice ever increasingly subtle evidence of a reaction occurring.

Some of our students come to us pretty well convinced that the chemistry of substances is quite mystical and mysterious and the pursuit of chemistry is the domain of magicians and sorcerers. These students are easy and defenseless targets for the fabricated, exaggerated "properties" conceived by advertising copywriters whose purposes are to make one brand of a commodity seem to be superior to competing brands. We stand to improve the consumer education of these students if they come to realize that the properties of substances really derive from the kinds, numbers, and arrangement of atoms making up molecules and the interactions among molecules.

In many respects, all events which our students see, smell, feel, taste, or hear can take on greater mystique if they associate events with the stretching, breaking, and reforming of bonds within or between molecules. But it is a kind of enlightened mystique. And this may well be the sort of thing that a chemistry course can contribute to the general education needed to live in an ever increasingly complex world.

INSTRUCTIONAL ACTIVITIES: STUDENT

3-1. Recognizing the Evidence of Chemical Changes. When substances are acted upon by forces or conditions that change their make-up, the action really takes place in the totally invisible realm of individual atoms and molecules. But we do see things happen—things happening due either to rearranging whole, unbroken molecules or to reshuffling fragments or even the individual atoms which make up molecules. If, when the forces or conditions which caused the reaction are exhausted or removed, the substance returns to its original make-up, this suggests that only physical changes took place, that is, whole, unbroken molecules had been rearranged.

If the substance does not return to its original make-up and a truly chemical event has occurred, the reshuffling of atoms or fragments of molecules is likely to produce such evidence as: (1) color changes, (2) the precipitation of insoluble substances by the interaction of initially soluble substances, (3) bubbles or smelly gases coming from initially solid or liquid substances, (4) temperature changes other than those caused by external heating or cooling, (5) redistributing the electrical charge balance between reacting species. Of course, continued study of chemistry will introduce you to additional, more subtle and sophisticated kinds of evidence.

Prepare a table listing the first four of the above kinds of evidence of chemical change. Postpone worrying about electrical effects, to observe this evidence calls for special equipment. Carry out each of the following reactions and record a "yes" or "no" if each reaction is or isn't accompanied by each kind of evidence. If you decide that only physical changes are involved, record "physical."

1. To about 5 ml of clear vinegar or 1 M acetic acid, add a pinch or two of baking powder or baking soda.

2. To about 5 ml of water, add *cautiously* a few flakes or crystals of a drain cleaning compound.

3. To about 5 ml of vinegar, add with even *greater caution* a few flakes or crystals of the same drain cleaning compound.

4. Using a hard glass test tube, heat a few crystals of paradichlorobenzene until the first bubbles appear. Keep the flame of the burner away from the mouth of the test tube.

5. Using a mortar, crush gently a mixture consisting of few crystals or lumps of ammonium sulfate and soda lime.

6. Using a 250 ml beaker, put a few drops of 6 M hydrochloric acid near one side and on the bottom of the beaker. Place a few drops of 6 M ammonium hydroxide opposite the acid. After a minute or so, combine the two.

7. Add a few drops of 6 M hydrochloric acid to 1 ml or so of 1 M sodium thiosulfate solution.

8. Add a few drops of 6 M ammonium hydroxide to 1 ml or so of 1 M copper sulfate solution.

9. With *caution,* add a drop at a time a few drops of 18 M sulfuric acid to about 10 ml of water in a large test tube. Save for reaction 11.

10. With *caution,* add a pellet of sodium hydroxide to about 10 ml of water in a large test tube. Observe until the pellet disappears. Save for reaction 11.

11. To each of the systems remaining from reactions 9 and 10, add a drop or two of phenolphthalein solution and record your observations as 11a and 11b. With *caution,* transfer half of the sodium hydroxide solution to another test tube and add, a drop at a time, several drops of the sulfuric acid solution. Record these observations as 11c. Now reverse this procedure by transferring half of the remaining sulfuric acid solution to another test tube and adding drops of the sodium hydroxide solution. Record observations as 11d.

12. Using a hard glass test tube, heat using *caution* about 1 ml of solid ammonium dichromate and remove the burner just as soon as you see evidence of changes occurring. Return the burner if the reaction seems to stop before all of the original substance has reacted.

13. Using a crucible, heat about 0.5 g of blue crystalline copper sulfate until no further changes seem to be taking place. As soon as it is cooled to room temperature, place the crucible in the palm of your hand and add water a drop at a time. Record observations as 13a and 13b.

14. To about 5 ml of commercial water glass, add a drop at a time a few drops of 6 M hydrochloric acid.

15. Using a hard glass test tube, heat about 1 ml of powdered lead dioxide. Assist your observations by inserting a glowing splint

from time to time during the heating. When no further changes can be seen, dump the contents into an evaporating dish and cool to room temperature. Record observations as 15a and 15b.

Doubtlessly, as you watched many of these reactions you saw things you would like to have explained. This will come. The important part for now is to be sure you stay alert while observing reactions and don't miss any of the clues that will help us develop good explanations. As the course unfolds, many of these reactions will be revisited and we will try to trace our observations down to the specific kinds of molecule or atom rearrangements from whence came the observations.

INSTRUCTIONAL ACTIVITIES: TEACHER

Sometimes our students seem to become impatient when they meet "exceptions" to what they hoped would be hard and fast "rules" or "laws." Things should be all black or all white, absolute rather than relative. Perhaps we foster this impatience in part by our use early in the course of reactions which "go all the way" or not at all. The following activity has been specially adapted toward loosening up student thinking.

3-T1. The Disappearing Precipitate. Announce that you will combine in all possible combinations, solutions of 0.5 M silver nitrate, 0.5 M sodium chloride, and 0.25 M sodium chromate. Included will be combinations of all three and the sequence with which the solutions are mixed is to be noted. Record all significant observations but avoid conversations with neighbors because the "game" we are playing pays off on the ability to observe effectively on one's own. There will be a special pay-off for spotting a particularly puzzling observation and the top prize goes to the person whose observations are good enough to let him explain this puzzling observation.

Each teacher, of course, can use his own finesse in confronting his students with the series of observations. The goal is to let them see that whereas both silver chloride and silver chromate form when silver ions encounter chloride and chromate ions simultaneously, the relatively greater solubility of the "insoluble" silver chromate allows it to be drained away to form silver chloride. An effective approach is to attach 9 burette clamps to a ring stand and use these to support test tubes. The three bottom test tubes can hold samples of the three original solutions for reference purposes. Each combination of solutions is kept on the ring stand until the total series is displayed. Assign a featured position to the test tube in which the silver ion solution is added to the already combined chloride and chromate solutions.

Again, assuming that large enough test tubes are used to enable all students to make their own observations, the teacher should discipline himself against wording the observations his students should record. Since you have to be doing something while the students are making and recording their observations, you might move up and down the aisles making sure that a maximum fraction of the students are catching the essential observations. This inspection activity also tips you off as to how flashy or how subtle each reaction should be arranged.

This activity can be paced so the dismissal bell will ring just when everyone's list of observations includes the "now it is and now it isn't" red-brown precipitate. This timing permits the next day's asignment to be the analysis of all observations and to come up with a plausible explanation. Except in rare situations, only a fraction of the members of a class will come up with a totally satisfactory explanation. For teachers who have a tendency to play games, rather than allowing the successful students to simply blurt out their explanations, the rules for the day's recitation might require all students to limit their contributions to asking cogent questions. Good questions will provide either straightforward or indirect reference to such factors as solubilities, solubility products, equilibrium, dynamic versus static systems, etc.

Audio-Visual Aids

3-AV1. "Combustion: An Introduction to Chemical Change." Color. 16 min. Iwamani Film.

3-AV2. "Physical and Chemical Change." Color. 28 min. Coronet. Contrasts the changes produced in salicylic acid and magnesium when subjected to physical versus chemical changes.

Ideas for Additional Instructional Activities

Although the reactions in Student Activity 3-1 bring out the basic observations which accompany chemical changes, you may prefer more spectacular albeit more complex reactions. Examples follow along with other ideas to extend the teaching of the topics presented in this chapter.

3-AD1. Moth balls placed in a tall container filled with a solution that is both 1 M sodium chloride and 1 M hydrochloric acid will show interesting effects if a few granules of zinc are added.

3-AD2. Place a watch glass containing 1 M silver nitrate on the stage of an overhead projector and then add a one cent piece.

3-AD3. Produce the patriotic colors by pouring 1 M ammonium hydroxide into 3 beakers containing phenolphthalein, lead acetate, and copper sulfate.

3-AD4. Prepare a solution containing 4 parts water and 1 part sodium silicate and then grow a "crystal garden" by dropping in crystals of salts known to produce colored ion solutions.

3-AD5. Place a gram or so of manganese dioxide in the bottom of a cylinder and then cover with a suds forming shampoo. Add 3% or stronger hydrogen peroxide.

3-AD6. Write with solutions of the following "invisible inks" and then develop as indicated: Silver nitrate with light. Lead acetate with hydrogen sulfide. Cobalt II nitrate with oxalic acid. Copper sulfate on blue paper with ammonium hydroxide.

3-AD7. With *caution,* add a few drops of glycerine to a small mound (0.5 g) of potassium permanganate.

3AD8. Carefully and *cautiously,* float 10 ml of ethanol on 10 ml of 18 M sulfuric acid. Sprinkle a few crystals of potassium permanganate through the two liquids. Flecks of fire will appear at the interface. Use *caution* when flushing contents down the sink.

3-AD9. This activity assumes knowledge of the hazards involved in working with yellow phosphorus and that adequate precautions against spontaneous ignition of the phosphorus can be maintained. Working under water, cut a pea-size quantity of yellow phosphorus. Shake dry and transfer promptly to a small beaker containing approximately 2 ml carbon disulfide. *Caution:* flammable and toxic. While the phosphorus is dissolving, arrange several graduated cylinders of 500 ml or greater capacity. Place two or three filter paper discs on the tops of the cylinders. When the reaction occurs, these paper discs may be ignited and tossed about. See Figure 3-1. For this reason, secure the discs to the cylinders using short lengths of thin wire and paper clips. Place a few drops, 6 to 10, of the phosphorus-carbon disulfide solution on each set of filter papers.

As the solvent evaporates, the dense carbon disulfide vapor produces an explosive mixture with the air in the cylinder. When the dissolved phosphorus is exposed to the air, it ignites. In turn, the explosive

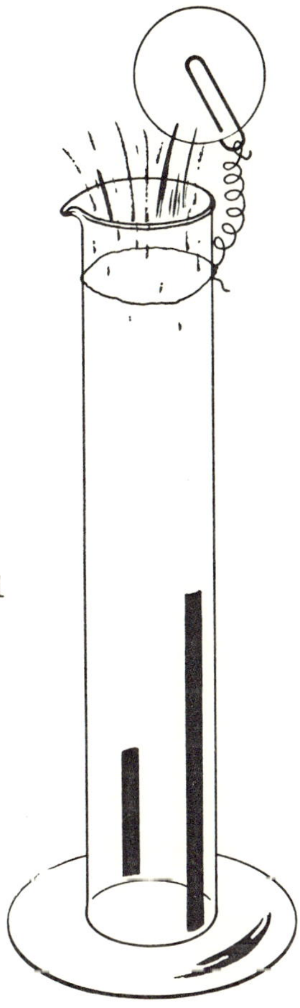

Figure 3-1

mixture in the cylinders is ignited. If different heights of cylinders are used, explosions of different pitch are produced, hence, the "barking dog" name for this demonstration.

This instructional activity yields a rich array of observations for those students who possess the required observational capabilities. Its explosively dramatic nature, however, invites many students to fail to catch many of the more significant observations. Teachers can well alert students to the large number, 15 to 20, separate events and cimcumstances which accompany the total series of events which culminates in the explosion. Additional enlightening observations occur if a second preparation of the phosphorus-carbon disulfide solution is prepared in which only one-fourth to one-half as much phosphorus is used to make the solution. Under these circumstances, the phosphorus merely smokes but does not burst into flame. What appears to be elemental sulfur is deposited on the filter papers, especially on the lower surfaces. These observations invite significant questions regarding the factors required to initiate chemical reactions.

3-AD10. Woodburn, John H., and Ellsworth S. Obourn, *Teaching the Pursuit of Science.* New York: Macmillan, 1965. A testing or examination activity based on the preceding instructional activity is described on pages 418–419.

3-AD11. Alyea, Hubert N., and Frederick B. Dutton, *Tested Demonstrations in Chemistry.* Easton, Pennsylvania: Journal of Chemical Education, 1960. A rich source of effective instructional activities dealing with many topics. Includes the 1955–56 series of "Tested Demonstrations in General Chemistry," the 1957–60 series of "Chem Ed Tested Demonstrations," and the "Demonstration Abstracts" presented in the *Journal of Chemical Education.*

3-AD12. Smith, Cyril S., "Materials," *Scientific American,* 217: 69. September, 1967. This is the introduction to the total issue which is devoted to the properties of substances, particularly, metals, ceramics, glasses, polymers, and composites. The article by Howard Reiss, "The Chemical Properties of Materials," on pages 211–220, is especially cogent.

4

THE PARTICLE NATURE OF MATTER

It is difficult to imagine students coming to a chemistry course without having heard of atoms and molecules. They have very probably been told repeatedly that all matter is composed of fantastically tiny particles and they may have been introduced to one or another concept of atomic structure. It is much less probable that our beginning students are able or accustomed to viewing chemical events from a particle perspective. To all too many students, baking soda is simply a white powder, vinegar is a pale brown or colorless liquid, and carbon dioxide is the fizzy, bubbly stuff that escapes from a bottled beverage. A chemical reaction is likely to be perceived as "some white stuff being poured from a balloon into a liquid causing the liquid to bubble and fill the balloon with some kind of gas."

The challenge we face is to give our students a functional, operational grasp of the significance of the particle nature of matter—to give them the wherewithal to build mental models of atoms and the structure of molecules. Supposedly, once they possess effective mental models of atomic structure, they will be able to bridge the gap between their observations of macroscopic features of a reaction and the submicroscopic domain in which the action really occurs. An ever-present hazard is that our students confuse their mental models with the reality they are intended to represent. Their models may take precedence over their observations—a state of confusion reflected in such questions as: "But wasn't each copper atom *supposed* to replace two silver ions?"

Contributing to our students' faith in the particle nature of matter is supposed to be the orderly relationships they "discover" in the relative numbers of particles taking part in chemical reactions. But this requires students to be able to count these particles. The dimensions of these particles, however, require us to use counting procedures which, to many of our students, are very indirect if not downright devious. In fact, few of the concepts we must teach will tax our ingenuity any more than the concept of relative atomic weights, the mole, and molar calculations.

INSTRUCTIONAL ACTIVITIES: STUDENT

4-1. The Idea of Relative Versus Actual Weights. Recall Activity 2-1. Reexamine your data and calculate the average weights of the small objects with which you worked (sequins and glitter stars). Use the platform balance and calculate the average weight of a third small object (beads). For example, another group of students found the average weights of sequins, glitter stars, and beads to be 0.005, 0.014, and 0.001 g. Do you see that the ratios of their weights are 5 : 14 : 1?

For the time being, disregard the actual average weights of the small objects. Think only of their relative average weights. Without counting as you go, weigh out two-tenths of the relative weight in grams of each object. In the above example, this would call for 1 g (5×0.2) of the sequins, 2.8 g of beads, and 0.2 g of stars. Now count the numbers of items in each of these weights. Don't be surprised if you find the same numbers of each of the three items. (That is, within the confidence limits of your balance.) This is the way relative weights work out.

Do you see the convenience in using the relative rather than actual weights when you want to count particles by weighing? The convenience would be much more obvious were you counting atoms and molecules, the fundamental albeit fantastically tiny particles with which chemists must work. The particles are so tiny that you would need twenty or so decimal places to show their individual actual weights whereas only a few integers are needed to show relative weights.

To add to your grasp of the usefulness of relative particle weights, weigh out 0.19 g of a mixture consisting of equal numbers of sequins and beads. Now weigh, without counting, 0.01 g of glitter stars. How does the number of glitter stars in 0.01 g compare with the number of sequin-bead combinations in 0.19 g of the mixture? Again, they should be the same; this is how relative weights work.

Do you see that equal multiples or fractions of the relative weight in grams of any particle contain the same number of items as there are in the same multiple or fraction of the relative weight in grams of any other particle? It is equally true that if you add the relative weights of two or more particles, there are the same number of combined units in a multiple or fraction of their combined relative weight in grams as there are in the same multiple or fraction of the combined relative weight in grams of any other combination of particles.

When we use balances to "count" particles by the relative weight method, we find ourselves using quite often the expression, "relative weight in grams." For convenience, let's abbreviate this to "relwig." (If you know that we are talking about the "mole," don't think we are playing funny games with you, we simply want everyone to understand its origin.) Show there are equal numbers of items in 0.4 relwig of sequins, beads, or glitter stars. Or in 0.4 relwig of sequin-bead, bead-star, or sequin-bead-star combinations.

By twisting things around a bit, we can also use the relwig as a kind of counting unit something like we use the dozen to count twelves of things. There are the same numbers of things in 1 relwig of any thing just as there are the same numbers of any thing in a dozen. Of the small objects you have been working with, how many items are in one relwig?

4-2. Using Relative Weights to Count Molecules. The substance, basic cupric carbonate, consists of particles composed of 2 copper, 1 carbon, 5 oxygen, and 2 hydrogen atoms. Charts are available which list the relative weights of all of the 100 or so different kinds of atoms, thanks to former generations of chemists. The relative weights of copper, carbon, oxygen, and hydrogen atoms are approximately 63.5 : 12 : 16 : 1. Thus, by addition, the relative weight of a cupric carbonate particle becomes approximately 221.

When basic cupric carbonate is heated to a sufficiently high temperature, it breaks down to release carbon dioxide and water vapor leaving a residue of copper oxide. The ratios of the relative weights of individual particles of these three substances are approximately 44 : 18 : 79.5. Hence, if cupric carbonate particles when heated lose one carbon dioxide molecule and one water molecule and are changed thereby to being two copper oxide molecules, the weights of these three products will always be in the ratios of 44 : 18 : 159.

To verify all of these "ifs," heat 2.21 g of the basic cupric carbonate in a preweighed crucible until no further changes seem to be occurring. Within the confidence limits of your balance, does the residue, after cooling, weigh 1.59 g? If not, your results disagree with the usual experience of chemists.

In the language of Activity 4-1, you used 0.01 relwig of the basic cupric carbonate. You drove off 0.01 relwig each of carbon dioxide and water molecules. You were left with 0.02 relwig of copper oxide, and 0.02 relwig of copper oxide should weigh 1.59 g. Rather than the nonsense unit, the *relwig,* the usual abbreviation for the "relative weight in grams" of any atom or molecule is abbreviated the *mole.* And rather than the 900 or so items in 1 relwig of sequins, glitter stars, or other make-believe atoms or molecules, one *mole* of particles actually contains approximately 6×10^{23} particles. This is why we can say that when we weigh out 2.21 grams of basic cupric carbonate (0.01 mole), we have approximately 6×10^{21} cupric carbonate particles. And the same logic enables us to say that, when heated, this weight of basic cupric carbonate lost 6×10^{21} molecules each of carbon dioxide and water and 12×10^{21} molecules of copper oxide were left behind.

And all of this is how the idea of relative atom and molecular weights enables us to count the particles that that are involved in the chemical reactions with which we work.

4-3. Realizing How Large a Number the Mole Is. Use a foot rule or caliper to estimate how many pages of your chemistry textbook are needed to give a layer one inch thick. There are 63,360 inches in a mile. How high would the stack reach if a mole of chemistry textbook pages were piled up? Would you be able to step off the top of the pile onto the moon?

Suppose your estimate of the number of textbook pages needed to make a one inch layer was too low by 2 pages and one of your classmate's estimate was too high by 2 pages. How high would the step be between the low and high estimates of a mole of chemistry textbook pages? A very, very sensitive balance may weigh gold no more accurately than plus or minus 1×10^{-10} mole. How high a stack of textbook pages would 1×10^{-10} mole of pages be?

INSTRUCTIONAL ACTIVITIES: TEACHER

4-T1. Man's Quest for an Effective Mental Model of the Atom. All in all, man's efforts to understand the structure of atoms add up to one of the most glorious of all of the episodes of science. And doubly so when we include how energy is involved in the structure of atoms and in their reshufflings from molecule to molecule. The total episode includes the contributions of thousands of scientists from all parts of the world. Some contributions were achieved centuries ago and although the first half of the twentieth century seems to mark the "atomic age," it may be many years before we are totally satisfied with the accuracy and precision of our concept of the structure of atoms and the forces within and between atoms and molecules.

The nature of this topic suggests we turn to the kinds of instructional activities more commonly used by social science teachers—activities intended to help students appreciate the interdependence of individual contributions. Sheer encyclopedic accounts fail to impress many of today's students. Sometimes this student apathy can be bypassed by framing assignments in one or another of the procedures which show through the work of many scientists. For example, students can be sent to reference materials in search of the original observations which might have sent a scientist on a particularly fruitful investigation; or to trace the sequence of background experiences and/or observations which led to a specific hypothesis; or to ferret out the strategy underlying the design of a particularly crucial experiment.

This approach leads to such assignments as: What kinds of experimental data added to Dalton's faith in his gradually evolving atomic theory? How might Faraday's publication of his investigations of "chemically equivalent weights" have contributed to Thomson's description of the electron? What hypothesis led Marie Curie in her search for what turned out to be radium? What evidence did Moseley use to back up his idea that the charge on the nucleus of an atom increases by regular steps as we pass from one element to the next element of higher atomic weight? What was the probable impact of the hypothesis that Albert Einstein worded in his 1939 letter to President Franklin D. Roosevelt?

4-T2. Interpreting Observations from the Perspective of the Particle Nature of Matter. Demonstration equipment is available which enables us to confront our students with some of the observations which puzzled scientists during the early phases of man's efforts to understand atomic structure and nuclear energy. Interesting lessons feature inviting our students to "explain" these observations as best they can with or without resorting to particle theory. Examples follow.

The green glow near the cathode of a discharge tube can be moved with a magnet (J. Plucker, 1862). Cathode rays travel in straight lines (H. Geissler, 1876) but because they can be moved or deflected by either an electric or magnetic field (H. Hertz, 1883), W. Crookes concluded in 1885 that cathode rays consist of streams of negatively charged particles.

Paper coated with barium platinocyanide glows brilliantly when near a discharge tube (W. C. Roentgen, 1895). Uranium salts emit radiation capable of penetrating opaque materials, exposing photographic film, and discharging electroscopes (H. A. Becquerel). Some of the radiation from uranium is deflected by a magnetic field and some is not (P. Curie, 1900).

4-T3. Visualizing the Mole as a Quantitative Unit. Chemists seem to acquire a degree of familiarity with the mole concept that makes it difficult to realize how students encounter so much difficulty in catching on to the concept and using it effectively and comfortably. The supreme importance of the concept, however, urges us to bend every effort to get the concept across.

Using a platform balance, weigh 63.5 g of pennies and announce that the students now see a mole of copper atoms. Unroll enough aluminum foil to weigh 27.0 g in order to exhibit a mole of aluminum atoms. Use iron nails, mercury, charcoal, sulfur, or other elements to extend the idea.

Weigh 87.9 g of iron II sulfide and place beside 55.8 g of iron and 32.1 g of sulfur. Announce that each sample contains one mole of particles, a mole of molecules in the case of the iron sulfide and a mole each of iron and sulfur atoms. Discuss the confusion that might arise from the expression that 1 mole plus 1 mole equals 1 mole.

Use graduated cylinders on platform balances to let the students see mole quantities of water, ethanol, carbon disulfide, and other liquids. Emphasize that although the samples of substances may contain equal numbers of particles at the molecular level, they may have different volumes and weights.

Fill large plastic bags estimated to hold 24.5 liters with various gases. Announce that you are exhibiting moles of gas molecules. State that each kind of gas, although occupying the same volume and having the same number of molecules, has its unique weight but pass up weighing the bags of gases. You may, of course, be clever enough to solve the logistic problems involved in determining the mass of this large a volume despite the complications due to the buoyancy of air.

Audio-Visual Aids

4-AV1. "Atomic and Molecular Models." Color. 32 frames, Britannica. Relationships between properties of molecules and their sizes, shapes, and structures.

4-AV2. "Building Atom Models." 3 min. Yale University. Features isomerism and steric hindrance.

4-AV3. "Construction of Molecular Models." Color. 32 frames. Britannica. Atomic, molecular, and ionic.

4-AV4. "Discharge Through Gases." 12 min. McGraw-Hill. Effects of progressive reduction of gas pressure.

4-AV5. "Electron Energy Level Chart." 42 × 29 inches. Welch Scientific. Energy levels represented by distance above baseline. Orbitals shown and occupancy can be shown in erasable chalk.

4-AV6. "Evidence for Molecules and Atoms." Color. 19 min. Britannica. Demonstrations, live-action photography of experiments.

4-AV7. "The Hydrogen Atom as Viewed by Quantum Mechanics." Chem Study. Demonstrates by analogy the meaning of atomic orbital; hummingbird collecting nectar.

4-AV8. "Light Sensitive Materials." Color. 22 min. Norwood Films. Physics of light and color. Composition of light sensitive substances.

4-AV9. "Molecular Model Kits." Edmund Scientific. One set using foam plastic and another using sponge rubber balls.

4-AV10. "Molecular Spectroscopy." Color. 23 min. Chem Study. Theory of light (infra-red) absorption and how used to identify molecules. Properties of light and interaction of light with matter.

4-AV11. "Molecular Theory of Matter." 11 min. Britannica. Development of and use to explain selected phenomena. Includes analogies.

4-AV12. "New Models of Molecules, Ions, and Crystals: Their Construction and General Use in Teaching Chemistry." Color. 44 min.

State University of Iowa. Describes construction of models to teach relationships between structures and properties and other fundamental principles of chemistry.

4-AV13. "A New Reality." Color. 51 min. Statens Filmcentral and Laterna Films, Denmark, and OECD, sponsored by the Carlsberg Foundation in Association with the International Council for Educational Films. Emphasizes the work of the Danish physicist, Niels Bohr.

4-AV14. "Orbitals—Atomic and Molecular." 44 frames. FOM. Sets of four quantum numbers to show the electronic configuration of the elements.

4-AV15. "Our Friend the Atom." Color. 48 min. Films. Historically oriented.

4-AV16. "What Makes Atoms Stick Together?" 30 min. Indiana University. Moves toward wave versus particle concept of atomic structure.

Ideas for Additional Instructional Activities

4-AD1. Alyea, Hubert M., "Photochemistry," *Journal of Chemical Education,* 44: A145. February, 1967. Demonstrations adaptable to projection.

4-AD2. King, Carroll L., and E. K. Neilsen, "Estimation of Avogadro's Number." *Selected Readings in General Chemistry,* Easton, Pennsylvania: Journal of Chemical Education, 1959. Pages 107–109. Based on determination of the thickness of a monomolecular film.

4-AD3. Ries, Herman E., "Monomolecular Films," *Scientific American,* 204: 152–164. March, 1961. Floating layers of molecules one molecule thick yield information on the molecules themselves.

4-AD4. Strong, Laurence E., and James Kreider, "The Copper-Sulfur System," *Current Topics* (Newsletter of CBA Project), Volume 5, No. 2, December, 1965. Detailed analysis of a reaction that is often used to confront students with observations intended to "prove" the law of definite composition on the basis of the particle nature of matter.

4-AD5. Wolthuis, Enno, Dale DeVries, and Marvin Poutsma, "Determination of the Equivalent Weight of Metals." *Suggested Readings in General Chemistry,* Easton, Pennsylvania: Journal of Chemical Education, 1959. Pages 112–113. A method giving good results for zinc, cadmium, and manganese.

4-AD6. Woodburn, John H., *Nuclear Science Teaching Aids and Activities,* United States Department of HEW, Washington, D. C., 1959. GPO. Pages 63–72. A listing in chronological order of selected observations, hypotheses, and experimental conclusions involving atomic structure and nuclear energy. Covers years between 550 B. C. and 1957.

5

COMMUNICATION IN CHEMISTRY

Some of our students are fascinated with the strange symbols and formulas of chemistry. Others may have become uneasy about the shorthand of chemistry through clumsy introduction to memory drills or premature assignments involving writing formulas or balancing equations. In either case, it is our goal to have all students come to look upon the language of chemistry as being an effective way to communicate.

The first major breakthrough happens when students associate the symbols of the elements with individual atoms and the formulas of compounds with individual molecules. It is very difficult to break the habit of thinking that Na stands for sodium or that NaCl stands for sodium chloride even though this habit seems to delay students from gaining the particle perspective from which to view chemical events.

Students meet similar difficulty in learning to read equations as descriptions of interactions among either individual or moles of atoms and molecules. It is a giant step from interpreting an equation as a description of the interaction between individual atoms, ions, or molecules rather than simply a statement of the initial and final components of the system.

Some of our students bring from their mathematics courses the ability to interpret the meanings of subscripts and coefficients as they are used in formulas and equations. Others do not and we must teach accordingly. The same can be said for such traditions as showing the reactants on the left and the products on the right of the arrow, listing the more positive particle first in a formula, using positive delta-H's to indicate endothermic reactions, and assigning zero as the oxidation potential for hydrogen and reporting other values relatively.

INSTRUCTIONAL ACTIVITIES: STUDENT

5-1. Symbols to Represent Atoms. Very early in the history of chemistry, people began using abbreviations and other short-cuts to make it easier to describe their discoveries and inventions. And when chemistry became a world-wide endeavor, the need became evident for a standardized "shorthand." While this need was developing, chemists throughout the world were becoming increasingly particle conscious. Thus, chemists began to agree that certain symbols would represent atoms of the elements.

Consult an alphabetical list of the elements and select the 10 elements whose names begin with letters closest to your own initials or "symbol." Use the section, "The Elements," in the *Handbook of Chemistry and Physics* or other references to explain how each of the 10 elements acquired its name and, hence, the symbol for its atoms. For example, be ready to tell why the name of a person or country is memorialized in the name of an element or what outstanding property of the element suggested its name.

(If this alphabetical assignment method leaves gaps, the missed items can be used for "make-up" or optional assignments.)

5-2. Formulas to Represent Substances or, More Precisely, Particles of Substances. Just as chemists found it convenient to use symbols to represent atoms of the elements, it became equally convenient to use formulas to show the kinds and numbers of each kind of atom making up a molecule of a substance.

Browse through your chemistry textbook and copy the names and formulas of 20 substances selected more or less randomly. What is the least number of different kinds of atoms to be found in a molecule? What substance in your list has the greatest number of atoms of all kinds in its molecules? Is there a minimum number of atoms needed to make a molecule? A maximum number?

Browse through your textbook again to see if you can find at least two substances whose names include each of the following prefixes or suffixes: -ic, -ous, -ide, -ate, per-, hypo-, hydro-, bi-, thio-, di-, tri-, tetra-, and -ite. After you have assembled your list, see if you can discover what common factor in the formulas of substances is designated by each of the prefixes and suffixes.

5-3. Translating Chemical Equations. Were you to visit one of the magnificient limestone caverns in our national parks, the guide might tell you that massive layers of limestone have been eroded by the action of carbon dioxide dissolved in water. In effect, the dilute solution of carbonic acid slowly changes the solid limestone to water soluble calcium bicarbonate. A chemist might summarize the guide's explanation in two equations:

$$H_2O + CO_2 \rightarrow 2H^+ + CO_3^=$$

$$2H^+ + CO_3^= + CaCO_3 \rightarrow Ca(HCO_3)_2.$$

The guide might also explain the beautiful stalagmites covering the floor of the cavern by saying that the water solution of calcium hydrogen carbonate dripping from the ceiling reverts to limestone, water, and gaseous carbon dioxide when it splashes on the cavern floor. Show that you are learning to use the shorthand of chemistry by translating the guide's explanation of stalagmites into an equation. There is a longstanding belief that atoms are neither created nor destroyed in ordinary chemical reactions. Have you kept this "law" in translating the guide's explanation? Is the cavern itself evidence that atoms of some kind have been destroyed?

Small quantities of oxygen can be produced conveniently by adding powdered manganese dioxide to a water solution of hydrogen peroxide. Water is a byproduct and for each mole of oxygen molecules produced, 46.92 kilocalories of heat are given off. Inasmuch as the manganese dioxide undergoes no permanent rearrangement of the atoms in its molecules, its function seems to be to catalyze the reaction. Be ready to explain how all of this information is communicated by:

$$H_2O_2 \xrightarrow{MnO_2} H_2O + \tfrac{1}{2}O_2 + 23.46 \text{ Kcal}$$

5-4. Making a Game of Learning Symbols and Formulas. Prepare a deck of 48 cards consisting of 4 cards each for the most commonly used elements or ions. It is wise to include 4 negative and 8 positive ions and to select those which have a variety of combining capacities. A suitable deck, for example, might include Na, Mg, Al, Sn, NH_4, As, Cr, and Fe for the positive ions and Br, S, PO_4, and SO_3 for the negative ions.

Individual game rules can be established as you go. One set of rules that seems to do nicely begins by dealing each player 5 cards and placing the remaining cards in the center of the table with one card exposed. Each player takes his turn by either drawing the exposed or top unexposed card. If he can produce a logical compound by using the cards in his hand, he can display the compound. If he does not name it correctly or if he cannot defend the combining capacities involved in the formula, he must return the cards to his hand and the player who recognized the goof receives game points. The player who goofed is also not allowed to discard at the end of his play whereas successful scorers complete the play by discarding. The game ends when the first player finds himself without any cards.

This game also lends itself to use with organic compounds. The hydrocarbon units take the place of positive ions and functional groups replace the negative ions.

INSTRUCTIONAL ACTIVITIES: TEACHER

The importance of the language of chemistry, especially chemical equations, cannot be overemphasized. But there is the hazard that our students can become caught up in the mechanical aspects of writing and balancing drill equations and lose sight of the true significance of what they are doing. The inevitability of conservation of atoms and of charge must also be emphasized in our instructional activities.

5-T1. Chemical Equations in Styrofoam. Foam plastic balls joined by short lengths of pipe cleaners provide effective linkage between the observable aspects of reactions and the mental models we hope our students envision ever more effectively as the course proceeds. Longer lengths of pipe cleaner can be used in the manner of ball and stick models whereas most foam plastics are soft enough to permit the models to be compressed in the manner of closely packed models. The relative volumes of atoms and ions can be suggested by choosing balls with appropriate diameters.

Each teacher easily finds the best way to use this instructional aid. A not too "hammed up" approach begins with permitting the students to observe a reaction the word equation for which is written on the chalk-board. A second set of glassware simulates the containers in which the reactants were stored and the reactant vessel. Rather than particles of actual atom or molecular size, however, these containers hold models of the particles. An example reaction is the action of water on sodium peroxide.

Begin the activity by taking one molecule of sodium peroxide and one molecule of water and hanging these models from one end of a balanced meter stick. From the other end, hang individual molecules of sodium hydroxide and oxygen and another ball to represent heat. The lack of balance is evident although in order to make the imbalance more dramatic, small lead fishing weights can be inserted in all of the plastic balls which simulate oxygen atoms. Achieve balance by taking additional molecules from the reactant containers. Emphasize that only whole molecules can be obtained on the reactant side and, hence, must be put into the reaction, (hung from the meter stick), as whole molecules.

Foam plastic balls can be colored conveniently by placing them in a container with dilute show-card paints, shaking thoroughly, and leaving to dry on absorbent paper. Some teachers prefer to print large symbols on the uncolored balls.

Pairs of balanced meter sticks can be used to simulate the half-reaction approach to balancing oxidation-reduction reaction equations. Whether to include or exclude spectator particles and how complex is the reaction to be modelled depends, of course, on the sophistication of your students at the time the instructional activity is being used.

Audio-Visual Aids

5-AV1. "The Chemical Formula." 35 frames. McGraw-Hill. Methods of writing formulas and naming simple compounds.

5-AV2. "Equations." 35 frames. McGraw-Hill. The information communicated by an equation and how to balance equations by inspection.

Ideas for Additional Instructional Activities

5-AD1. "The Elements—Materials for a Growing World of Knowledge," *Chemistry,* 38: 12–15. November, 1965. Table listing the elements, their symbols, atomic numbers, atomic mass, discoverer and date, and derivation of names.

5-AD2. Copley, George, "Linear Algebra of Chemical Formulas and Equations," *Chemistry,* 41: 22–27. October, 1968.

5-AD3. Fabishak, Vernon L., "Matrix Method for Balancing Chemical Equations," *Chemistry,* 40: 18–21. December, 1967.

5-AD4. Shen, Samuel Y., -Huang, Suei-Rong, and Stewart Karp, "The Significance of a Simple Equation," *Chemistry,* 42: Pages 16–18. February, 1969.

5-AD5. Van Tamelen, Eugene E., "Benzene—The Story of Its Formulas 1865–1965," *Chemistry,* 38: 6–11. January, 1965. How the formula for benzene became the starting point for numerous questions and discoveries.

5-AD6. Walker, Ruth A., and Helen Johnston, *The Language of Chemistry: Reading, Writing and Understanding Equations.* Englewood Cliffs, New Jersey: Prentice-Hall, 1967.

5-AD7. Wilchers, Edward, "Why the Carbon-12 Scale," *Chemistry,* 37: 12–17. March, 1964. Reasons for changes in atomic weight scales.

6

STOICHIOMETRY

Among many of today's young people, chemistry has the reputation for being a difficult, demanding course. Fear of the arithmetic of chemistry doubtlessly contributes to this reputation. The origins of this fear and its effects on student attitudes are obscure. But the uneasiness it seems to generate must be kept in mind when we plan instructional activities involving stoichiometry.

In general, today's students have been taught all of the mathematics they need to handle the quantitative descriptions they encounter in high school chemistry. But several years may have passed since they were last required to retrieve and apply portions of their mathematics. In fact, many students seem to have had little experience in recognizing situations in which certain mathematical skills apply even though they may have mastered these skills in prior courses.

One of the greatest values in the chemistry course for general education students derives from the opportunity it provides to review computational skills and to shed additional light on mathematics as another highly significant human endeavor.

INSTRUCTIONAL ACTIVITIES: STUDENT

6-1. Equations as "Recipes" for Chemical Reactions. Several of the working ideas of chemistry can be used to obtain a statement of the quantities of each reactant needed to produce the desired quantity of any product. We know, for example, that no atoms will be lost or created during a chemical reaction. The idea of relative atomic weights allows us to weigh out known numbers of each kind of atom or molecule we put into the reaction and, if we have a properly balanced equation, predict the numbers of each kind of atom or molecule that is produced by the reaction. In other words, a proper equation is a "recipe" for the reaction, assuming that the reaction goes as we predict it will and produces a 100% yield of the predicted products.

Consider, for example, that very important reaction in which green plants combine carbon dioxide and water to produce glucose, oxygen, and water. The total process can be summarized by the equation:

$$6CO_{2(g)} + 12H_2O_{(l)} \xrightarrow[\text{light energy}]{\text{chlorophyll}} C_6H_{12}O_{6(aq)} + 6O_{2(g)} + 6H_2O_{(g)}$$

This "recipe" can be read as: Six portions of carbon dioxide combined with twelve portions of water in the presence of chlorophyll and sunlight yields one portion of glucose, six portions of oxygen, and 6 portions of water.

The idea of relative atomic weights allows us to use a balance to measure out the "ingredients." Read from this point of view, the "recipe" becomes: 264 grams (6 x 44g/mole) of carbon dioxide combined with 216 grams (12 x 18g/mole) of water yields 180 grams of glucose, 192 grams (6 x 32g/mole) of oxygen, and 108 grams (6 x 18g/mole) of water.

Just as kitchen recipes can be adjusted to accommodate large or small families, chemical "recipes" can be adjusted by simple ratio and proportion. For example, can you see that 1920 grams of oxygen would be produced if a field of corn used 2640 grams of carbon dioxide in the photosynthesis reaction? Or that if a single leaf produced only 0.018 gram of glucose, it would remove 0.0264 gram of carbon dioxide from the atmosphere?

For a second example, the burning of octane, one of the possible components of gasoline, can be shown by the equation:

$$C_8H_{18(l)} + 12\tfrac{1}{2}O_{2(g)} \rightarrow 8CO_{2(g)} + 9H_2O_{(g)} + 1317.5\text{ Kcal}$$

Using this equation as a "recipe," do you see that an automobile which uses 114 grams of octane to go a mile changes 400 grams of atmospheric oxygen to carbon dioxide? Can you use this same "recipe" to calculate how much carbon dioxide would be released along a one mile stretch of highway if you know how many cars traverse the mile in a specified unit of time?

6-2. Volumes as Well as Weights from Chemical Equations. Some substances, especially gaseous substances, are more easily measured by volume than by weight. A chemical equation can also be interpreted in volume units in so far as gases are concerned. By definition, a gas completely fills its container. On this basis, equal volumes of all kinds of gases at the same pressure and temperature contain equal numbers of molecules. Thus, by reading the equation in the language of moles, we obtain the relative volumes of each gaseous substance in the "recipe."

For example, recall from 6-1 the equation for burning octane. Look at the three gaseous substances in the reaction. Do you see that for each 12½ volumes of oxygen used to burn the octane, 8 volumes of carbon dioxide and 9 volumes of water vapor are produced? It has been determined that a mole of gas particles occupies approximately 24.5 liters at 25°C. and one atmosphere of pressure. We can use this conversion factor to say that when 114 grams of octane (weight of one mole) are burned, 286.25 liters (12½ x 24.5 liters/mole) of oxygen are consumed and 200 liters (8 x 24.5 liters/mole) of carbon dioxide are produced.

For a second example based on the photosynthesis equation in 6-1, do you see how we can say that green plants exchange equal volumes of oxygen for carbon dioxide during the synthesis of glucose? Or that for every 216 ml of rainwater (12 x 18g/mole x 1 ml/g) that finds its way into the tissues of the green plant, 147 liters of water vapor would be added to the atmosphere?

If we know the density of a substance that is involved in a chemical equation, even though the substance isn't a gas, we can calculate the volume of any specified weight. The density of octane, for example, is 0.7 g/ml. Thus, 114 grams would occupy approximately 163 ml or 0.163 liter. In everyday language, an automobile going one mile burns about a teacup of gasoline and in doing so converts the oxygen in 370 gallons of air to 50 gallons of carbon dioxide and adds 1317.5 kilocalories of heat to the environment.

6-3. Determining the "Recipe" When You Can Analyze the Product. In Activities 6-1 and 6-2, you saw how the weights or volumes of reactants and products can be calculated when the equation for the reaction is known. Can you see that it would be equally possible to determine the equation for a reaction when the weights or volumes of the reactants and products are known?

Suppose we know that when hydrogen burns in oxygen water is produced and that we have a laboratory set-up that shows us for each two volumes of hydrogen burned, one volume of oxygen is used and two volumes of water vapor are produced. The equation would be:

$$2H_{2(g)} + O_{2(g)} \rightarrow 2H_2O_{(g)}$$

On this basis, what would be the equation for producing ammonia from nitrogen and hydrogen if you found that for each volume of nitrogen, 3 volumes of hydrogen were consumed and 2 volumes of ammonia were produced?

Let's look at the burning of a substance similar to octane from the point of view of not knowing the equation for the reaction but knowing the weights or volumes of the substances involved in burning a sample. In this case, when a 100 gram sample of substance "X" was burned, 352 grams of oxygen were used and 308 grams of carbon dioxide and 144 grams of water were produced. In the language of portions, this means that 11 portions or moles (352 grams ÷ 32 grams/mole) of oxygen were used. Similarly, 7 portions or moles (308 grams ÷ 44 grams/mole) of carbon dioxide and 8 portions or moles (144 grams ÷ 18 grams/mole) of water were produced. Thus, the equation becomes:

$$X + 11O_{2(g)} \rightarrow 7CO_{2(g)} + 8H_2O_{(g)}$$

And a little additional thinking will reveal the composition of substance "X." The 100 gram sample must have contained 7 portions or moles of carbon atoms and 16 portions or moles of hydrogen atoms. This means its formula is C_7H_{16}, a fact further verified by noticing that the weight of one mole of this substance, heptane, is 100 grams.

6-4. Comparing Actual with Theoretical Yields. Golfers seem to enjoy finding out if they can go around a golf course in par, that is, use no more strokes than are used by an expert golfer. You may enjoy an "against par" type of laboratory exercise. The "game" is to see if your yield from a chemical reaction corresponds to the yield a "perfect" chemist would obtain. But the chemistry "game" differs from the golfer's "against par" game. For one reason, the perfect yield from a chemical reaction is controlled by the law of conservation of atoms. The total weight of all substances after the reaction is over always equals their initial weight.

But you can never be sure what the products are when you carry out a chemical reaction and you may allow something to escape because its properties differed from what you expected. Nor can you be totally sure that all of a solid precipitate didn't pass through the filter paper, that all substances were equally dry when you weighed them, or that your data haven't been compromised by similar procedure errors.

For one such game, plan to convert silver to silver nitrate and thence to silver chloride. The equation for these reactions suggests that one mole of silver yields one mole of silver nitrate and that one mole of silver nitrate, in turn, when treated with sodium chloride, yields one mole of silver chloride.

$$Ag \xrightarrow{HNO_3} AgNO_3 \xrightarrow{NaCl} AgCl$$

For sake of economy and convenience, use only 1/50th of the basic or mole "recipe," that is, begin the exercise with only 0.02 mole (2.16 g) of silver. This means, according to "par" for this reaction, you should produce yields of 0.02 mole (3.40 g) of silver nitrate and 0.02 mole (2.87 g) of silver chloride.

Convert the silver to silver nitrate by placing the silver in a carefully weighed 150 ml beaker and adding 10 ml of 6 M nitric acid (*fume hood*). If the reaction is too slow to show brown vapors coming from the liquid phase, warm gently. After the reaction is complete, continue to heat gently to evaporate the liquid phase and to drive off the nitric acid not used to produce the silver nitrate. However, allow the silver nitrate crystals to form slowly by leaving the almost evaporated residue stand for a day or so. Dry the residue in an oven or dessicator, weigh, and see how closely your yield agrees with the anticipated 0.02 mole weight.

For the second step, dissolve the silver nitrate crystals in about 10 ml of distilled water. In a second beaker, dissolve about 3 grams of sodium chloride in about 10 ml of water and then add this solution to the silver nitrate solution. Do you see that this weight of sodium chloride insures that all of the silver will be precipitated as silver chloride? Boil the resulting system for a few minutes and then filter off the yield of silver chloride. Dry and weigh the yield of silver chloride to see how closely your yield agrees with the anticipated 0.02 mole weight. Which of your two yields, the silver nitrate or the silver chloride, agrees most closely with "par?" Why? Where could errors most likely have creeped into your work?

6-5. Comparing Actual with Theoretical Yields When the Reactants Are in Solutions. In many chemistry laboratories, it is more convenient to dissolve certain reagents and keep the solutions on hand to be used as needed. For example, a full mole of silver nitrate (169.87 g) can be dissolved in enough water to make one liter of solution. When small quantities of silver nitrate are called for, each milliliter of this solution contains 0.17 gram of silver nitrate or, in other words, 0.001 mole. Similarly, each milliliter of the solution contains 0.0010 mole of dissolved silver ion as well as 0.0010 mole of nitrate ion.

For further practice in the "against par" type of chemistry game, plan to prepare silver chromate, lead chromate, lead iodide, and/or lead bromide. Find out what concentrations of the required reactant solutions are available in your laboratory and calculate the volumes you would need to prepare a few grams of the desired product. Write accurate equations for the reactions used to produce the desired product else you will be without "recipes" for your reactions. When you compare your yields with predicted or theoretical yields, you may want to check the *Handbook of Chemistry and Physics,* Section B, Physical Constants of Inorganic Compounds, to adjust your data for the solubility of the product. In the case of lead bromide, for example, approximately one gram of this salt will dissolve in 100 ml of water at room temperature.

Although a bit more complicated, the reaction between hydrochloric acid and baking soda provides a good stoichiometric game. When the resulting solution from this reaction is evaporated, assuming an excess of the acid was used, sodium chloride remains as the only solid product. Write the equation for this reaction. Decide what would be an appropriate weight of soda to use and then calculate the predicted yield of sodium chloride and the amount of hydrochloric acid you will need to insure a complete reaction. After drying your yield of salt crystals, compare the weight with anticipated results. If there is a discrepancy in these weights, when you seek an explanation, keep in mind the hygroscopic nature of sodium chloride.

6-6. Using Stoichiometry to Find the Correct "Recipe." Examine crystals of copper sulfate and ammonium sulfate. If these two substances are dissolved and their solutions mixed in the proper proportions, a new substance will be produced which, when it crystallizes from solution, forms uniformly colored crystals. Assuming you do not know the formula of the product, what weights of the two reactants, copper sulfate and ammonium sulfate, would be logical quantities to use in an experiment to determine the formula of their product? Keep in mind that the strategy here is to bring together the weights of the two reactants that will result in a crystallized product that is not contaminated with an excess of either reactant.

Discuss your hypothesis with those of several classmates. Test your hypotheses by weighing out the predicted weights of the reactants, prepare solutions, mix the solutions, and then evaporate the resulting system slowly to allow the product (or products) to crystallize. Examine the resulting crystals and decide which "recipe" is the most accurate description of the reaction between copper sulfate and ammonium sulfate.

6-7. Using Stoichiometry to Solve a Chemist's Problem. From a legal point of view, sterling silver contains 92.5 per cent silver and 7.5 per cent copper. Stoichiometry can be used to prove that something labelled "sterling" silver contains the full 92.5 per cent silver. The logic is to dissolve the silver in nitric acid and then add a reagent that reacts with the silver by precipitating it from the solution. By knowing the amount of the reagent used to precipitate the silver, we can calculate the weight of silver originally present in the solution.

Weigh very carefully approximately 1 gram of a silver alloy that is supposed to be "sterling" silver. Dissolve it in a minimum amount of 6 M nitric acid. Dilute the resulting solution to about 50 ml with distilled water. Add 5 ml of 0.1 M ferric nitrate and filter off any precipitate that appears. Discard this precipitate.

Prepare one liter of 0.1 M potassium thiocyanate solution. The thiocyanate ion reacts with silver ion to produce a white precipitate of silver thiocyanate but when all of the silver has been precipitated from the solution, the thiocyanate ion reacts with the ferric ion to produce a very highly colored thiocyanoiron ion. It is the appearance of this dark red ion that will tell you when to stop adding the thiocyanate solution.

Fill a thoroughly cleaned burette with the potassium thiocyanate solution. Record the exact volume of this solution required to precipitate the silver from your "sterling" silver sample. Assume that each mole of the thiocyanate ion precipitates a mole of silver and, knowing that each ml of the thiocyanate solution contains 1×10^{-4} mole of the thiocyanate ion, calculate the amount of silver in your solution. If this equals 92.5 per cent of the original weight of your metal sample, the metal meets the requirements to be sterling silver.

INSTRUCTIONAL ACTIVITIES: TEACHER

6-T1. Discovering the Value of the Coefficients in a Chemical Equation. Laboratory exercises in which students are to "discover" the ratios between moles or numbers of molecules of reactants and products often "missfire" with many students. The good student who knows the significance of a formula wonders why he can't simply balance the equation in order to determine the proper coefficients. The inept student's weight data seldom point to clear-cut mole ratios. Furthermore, these exercises come sufficiently early in the course that many students are so occupied with the sheer mechanics of the reaction that they may fail to recognize the central thrust of the exercise. For these reasons, many teachers prefer to back up individual laboratory exercises involving mole ratios with at least a demonstration or two.

Prepare 0.1 M solutions of sodium chloride, calcium chloride, aluminum chloride, and stannic chloride. Place 10 ml of each solution in each of 4 test tubes. Add to each test tube enough 0.5 M silver nitrate to precipitate all chloride ion as silver chloride. Although the silver chloride may not settle out equally promptly in each of the test tubes, the students obtain prompt evidence of the approximate ratios of the yields of silver chloride. Plan to filter off the silver chloride from each test tube and while waiting for the precipitate to dry, indicate the probable coefficients for each of the four reactions based on estimated yields.

This type of exercise may disturb some students. If the student has missed the significance of molarity as a way to express the concentration of a solution, the logic of the whole exercise breaks down. If he understands molarity, the exercise can only confirm his expectations. It follows that the teacher must be prepared to present the exercise in such a way as to invite "discovery" but, at the same time, avoid putting students into a situation in which they must feign something they really do not feel or believe in.

6-T2. Taking Students Through a "Weight-Volume" Problem. Weigh a zinc-coated nail as accurately as possible. Fill a 100 ml graduated cylinder with 1.0 M hydrochloric acid. Put about 600 ml of water in a liter beaker. Have an assistant use crucible tongs to hold the nail standing on its head on the bottom of the beaker. Cover the graduated cylinder with a square of paper and invert the cylinder over the nail.

On the following day, read the graduate to determine the volume of hydrogen that has been replaced by the zinc from the nail. (It is wise to carry out preliminary investigations to make sure that the nails you are using do not generate more hydrogen than the graduated cylinder you plan to use will hold.) Correct the volume for partial water vapor pressure. If appropriate to do so at this stage of the course, also correct the pressure and temperature to STP, otherwise assume that a mole of hydrogen gas occupies 24,500 ml under the usual laboratory conditions.

Use the equation for the reaction to calculate the weight of zinc that must have been lost by the nail before removing the nail from the beaker. After setting the stage of anticipation to see if the actual weight loss will agree with the "theoretical," remove the nail, dry, and weigh.

Anticipate questions about why the iron of the nail is not attacked by the acid with subsequent additional yield of hydrogen. Such questions may well be used to arouse additional investigations but it is well to be prepared to test for the presence of ferric ions in the solution remaining in the beaker.

Audio-Visual Aids

6-AV1. "Atomic Weight Determination." 17 min. Purdue University. Using oxides to calculate atomic weights.

6-AV2. "Determining Molecular Formulas." Color. 12 min. McGraw-Hill. Using percentage composition and Avogadro's Hypothesis.

6-AV3. "The Mole Concept." 32 frames. Britannica. The introductory frames treat the mole concept and its relation to the Avogadro number.

6-AV4. "Molecular Weight of Oxygen." Color. 11 min. Coronet. Stoichiometrically from heating potassium chlorate.

6-AV5. "Model for Weight Relations in Chemical Reactions." 3 min. Yale University. Molar relations.

6-AV6. "The Molar Method, Chemistry Concepts." Programmed Instructional Materials. 309 frames. Reusable book. Mole method of solving a variety of problems, e.g., percentage weight composition.

6-AV7. "Volumetric Flasks, Chemical Techniques." Color. 4 min. Norwood Films. Proper use of in preparing standard solutions.

Ideas for Additional Instructional Activities

6-AD1. Carmody, Walter R., "The Determination of the Combining Weight of Tin," *Journal of Chemical Education,* 44: 416. July, 1967. An analysis of this "standard" exercise.

6-AD2. Child, W. C., Jr., and R. W. Ramette, "The Stoichiometry of an Oxidation-Reduction Reaction," *Journal of Chemical Education,* 44: 109–110. February, 1967. Determination of nitrogen as a product.

6-AD3. Cohen, Irwin, "Moles and Equivalents: Quantities of Matter," *Journal of Chemical Education,* 38: 555–556. November, 1961. Computations involving the mole concept.

6-AD4. Copley, George Novello, "The Mole in Quantitative Chemistry," *Journal of Chemical Education,* 38: 551–553. November, 1961. Applications of.

6-AD5. Guggenheim, E. A., "Molecules Versus Moles," *Journal of Chemical Education,* 43: 250–251. May, 1966. Use of the term, "gram-molecule" versus, "mole."

6-AD6. Dingledy, David, and Walther M. Barnard, "The Stoichiometry of Sulfides," *Journal of Chemical Education,* 44: 693–694. November, 1967. Analysis of this "standard" exercise.

6-AD7. Henry, Paul S., "Evaluation of Avogadro's Number. A General Chemistry Experiment," *Journal of Chemical Education,* 43: 251–252. May, 1966. Based on determination of average particle population in a suspension as a function of height.

6-AD8. Lee, Shiu, "A Redefinition of 'Mole'," *Journal of Chemical Education,* 38: 549–551. November, 1961. Proposes the term, "gram formula weights."

6-AD9. Scholes, S. R., "On Avogadro's Number," *Journal of Chemical Education,* 42: 126. March, 1965. A poetic reflection.

6-AD10. Theiman, Theo, "Determining the Empirical Formula of Sodium Hydroxide," *Chemistry,* 38: 32–33. July, 1965. By reacting known mass of sodium with water and measuring volume of hydrogen produced.

6-AD11. Thiessen, G. W., "Finding Atomic Weights," *The Science Teacher,* 22: 140. April, 1955.

6-AD12. Yingling, R. T., and H. A. Neidig, "Stoichiometric Relationships in Two Systems," *Chemistry,* 37: 24–25. April, 1964. *Chemistry,* 40: 34–37. January, 1967. *Chemistry,* 38: 24–29. February, 1965.

Copper (II) nitrate-ammonia and mercury (II) nitrate-potassium iodide. Describes the continuous variation design for investigating stoichiometric relationships.

7

PROPERTIES OF GASES

Although students seem to have a common sense understanding of the relationships between the volume, temperature, and pressure of a specified mass of gas, they tend to have trouble with the arithmetic that enables them to apply the gas laws. In fact, experienced teachers identify the properties of gases as one of the topics wherein students are likely to become so caught up in computations that they lose sight of the principles or concepts to be served by the computations.

In most cases, it is in the estimation of the number of particles in a measured volume of a substance in the gaseous state that students tangle with the gas laws. On this basis, some teachers prefer to postpone the refinements inherent in the "ideal gas" concept, and operate on the assumption that a mole of gas particles occupy 22.4 liters under standard or 24.5 liters under room conditions (25°C. and 760 mm of Hg). The Avogadro Hypothesis and the law of partial pressures are also represented as clear-cut information which students can apply.

Because students have very probably been introduced to Boyle's, Charles', and Graham's laws in general science courses, they may question the value of instructional activities designed to "discover" or "prove" these laws unless the activities have unique features. Fortunately, the gas laws have generated a great many interesting and effective instructional activities, especially lecture demonstrations.

INSTRUCTIONAL ACTIVITIES: STUDENT

7-1. Relationship Between Volume and Pressure of a Trapped Quantity of Gas. Here is an experiment performed many years ago by a famous English physicist. See if you can redo the experiment by taking advantage of this description of the experiment performed over 300 years ago by Robert Boyle at Oxford, England.

Some experiments being done by our English, French, and Italian friends suggest that the greater the weight is that leans upon the air, the more forcible is the endeavor of dilation, and consequently its power of resistance (as other springs are stronger when bent by greater weights). Today, I want to tell you about an experiment to test this.

Take a long glass tube and by a dexterous hand and the help of a lamp, crook the tube at the bottom so that the part turned up is almost parallel with the rest of the tube. Seal the orfice of this shorter length of the siphon (if I may so call the whole instrument). Divide the length of it into inches and divide each of these into eight parts by a straight strip of paper carefully pasted all along it.

Put in as much quicksilver as needed to fill the arch or bended part of the siphon so that the mercury standing at a level might reach in the one leg to the bottom of the divided paper and adjust to the same height of horizontal line in the other. Take care by frequently inclining the tube to see that the air might readily pass from one leg into the other by the sides of the mercury so that the air at last included in the shorter cylinder is of the same laxity with the rest of the air about it. (If you are not as dexterous as Mr. Boyle, you may prefer to leave the glass tube unsealed until after you have placed enough mercury in the bent tube to seal off both sides of the U.)

Pour quicksilver into the longer leg of the siphon until the air in the shorter leg is by condensation reduced to take up by half the space it possessed before. Cast your eyes upon the longer leg of the glass and observe not without delight and satisfaction that the quicksilver in that longer part of the tube is about 29 inches higher than the other.

Can you go Mr. Boyle one better and prove that the product

of the volume of the trapped air (length of tube) and the pressure it is under (initial barometric reading plus the difference in heights of the two mercury columns) produces a constant?

Note: Because students are prone to spill mercury, safety precautions involving reducing exposure to mercury vapor suggest that this instructional activity should be presented as a demonstration.

7-2. Relationship Between Volume and Temperature of a Trapped Quantity of a Gas. Fit a dry 250 ml Erlenmeyer flask with a one-hole rubber stopper carrying a 4 cm length of glass tubing. Immerse this flask in a large beaker of water using a clamp attached to a ring stand to keep all of the flask immersed except the glass tubing. Heat the water in the beaker and assume that the air in the flask is heated to the same temperature.

As soon as the water in the beaker is boiling vigorously, record the water temperature and quickly remove the flask. Hold your finger over the end of the glass tube until you have inverted the flask and submerged it in a container of ice water. Remove your finger and allow as much water as will to enter the flask. Assume that the temperature of the air remaining in the flask is the same as the surrounding cold water.

Pour the water from the flask into a graduated cylinder and subtract this volume from the quantity of water needed to fill the flask. Thus you obtain the volume of a trapped mass of air and the changes in volume that it underwent while cooling from the temperature of boiling to the temperature of cold water. Plot these two temperature-volume relationships with the volumes as ordinates and temperatures as abscissas. Extrapolate to determine the temperature to which the air would have to be cooled to reduce the volume to zero. How close does your value compare to the generally accepted value of $-273°C$? This is 0° Kelvin, a temperature scale with a different zero base but with each degree equal to a Celsius degree. According to your data, which of the following relationships is most nearly true: $P \times T$ (degrees Celsius) equals a constant. $P \times T$ (degrees Kelvin) equals a constant. $P \div T$ (degrees Celsius) equals a constant. $P \div T$ (degrees Kelvin) equals a constant.

7-3. The Volume Occupied by a Mole of Gas Particles. If it is true that equal volumes of gases contain equal numbers of particles then we can argue that equal numbers of gas particles should occupy equal volumes. This line of reasoning leads us to the conclusion that there is a definite volume we can expect a mole of any kind of gas molecules to occupy.

To verify this idea, weigh to hundredths of a gram a container of butane or other bottled gas fuel of known composition. Attach a rubber hose to the nozzle and collect about 85 ml of the gas by water displacement. Remove the rubber hose and reweigh the gas container. Determine the weight of gas used and scale up the volume it occupied to obtain the volume that a full mole of the gas would have occupied. Remember that when gases are collected by water displacement, they are likely to be "contaminated" with water vapor. Correct for this "contamination" by subtracting the water vapor pressure for the temperature at which the gas was collected from the barometer reading and then adjust the volume collected by multiplying this volume by the fraction:

$$\frac{\text{Pressure of the gas without water vapor pressure}}{\text{Barometric reading}}$$

Assuming a room temperature of 25°C and a barometer reading of 760 mm of mercury, a mole of gas particles occupy 24.5 liters. How close to this value did you come?

INSTRUCTIONAL ACTIVITIES: TEACHER

7-T1. The Displacement Rates of Gases. The vaporization of liquids and the subsequent diffusion of the substances in the gaseous state, at first thought, appears to be a highly random event. In fact, one of the objectives underlying this instructional activity may be achieved better if its introduction features several questions which emphasize the random aspects of the phenomena that are involved. For example, if a boy is sitting exactly equal distances from two girls each of whom is wearing a unique kind of perfume, will both fragrances reach the boy at the same moment? And what assumptions must be applied to various variables in order to attempt to answer such a question?

The diffusion of hydrogen chloride and ammonia provide a classical activity for studying the displacement rates of gases. Small quantities of maximum concentration hydrochloric acid and ammonia water are placed at opposite ends of a glass tube and the appearance of a white ring of ammonia chloride marks the meeting place of the advancing vapors. Glass tubes approximately 1 cm in diameter and 100 cm long serve nicely. The reagents may be soaked in cotton plugs, taken up in cotton embedded in corks, or placed in combustion boats and inserted simultaneously into the ends of the tube. Anticipate that the hydrogen chloride will advance approximately 0.7 and the ammonia 1.07 mm per second.

To challenge your students to recognize orderly relationships in the diffusion rates of gases without giving them helpful hints may lead to frustration. You may want to provide displacement rates for additional gases in the hope that the relationship might become more easily identified. Some such data are: hydrogen 30, helium 21, nitrogen 8.0, dimethylamine 6.4, trimethylamine 5.5 cm/100 sec.

Additional hints may call attention to molecular weights and the subsequent differences in the masses of the particles being diffused. The subtlety of hints regarding mathematical analysis will depend upon the backgrounds of your students but few students will discover the anticipated relationship unless they attempt to graph displacement rates against molecular masses. Logarithmic plots provide effective hints of the fact that diffusion rates are proportional to the inverse of the square root of the molecular mass.

7-T2. The Properties of Gases and the "Thirsty" Bird. Although the "drinking bird" toy (Figure 7-1) is well known, surprisingly few students have come upon explanations of its action—action involving several of the principles and concepts dealing with the properties of gases. Evaporation of water from the cloth covering of the "head" end cools the gas in that end of the inner tube which, in turn, causes the greater pressure in the bulb at the "tail" and thus driving the entrapped liquid past the balance point. But as the tube approaches the horizontal, the "seal" between the two bulbs is broken and the liquid flows back toward the "tail" end. A special feature of this visual aid is the way it links the kinetics of gas properties with an obvious display of energy transformation.

If the "thirsty bird" is allowed to "drink" in a display case for a day or so in advance of opening the topic of the behavior of gases, student curiosity is whetted. Especially so if the teacher asks occasional questions such as: I wonder if there is any way to cause the bird to drink more often? What do you think would happen if we add a little ethanol to the water?

7-T3. The Vapor Pressure of Liquids. Fit three bell jars with

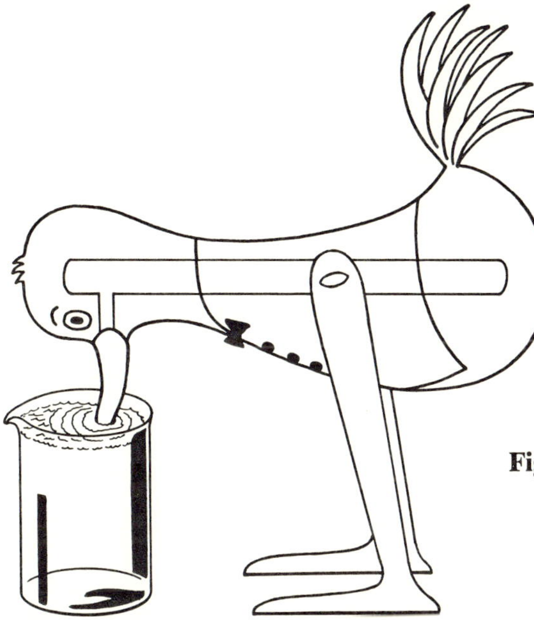

Figure 7-1

manometer tubes. Place equal weights of hot, cold, and room temperature water in similar beakers and place on large glass plates. Cover each beaker with a jar making sure that there is an airtight seal between the jar and the glass. After observing the manometer tubes for several minutes, replace the three beakers of water with beakers containing equal weights of water, ethanol, and ether. Announce that the three liquids are at the same temperature and invite predictions of how the manometers will read after a few minutes. As the end of the day's session approaches, replace the beakers, one with saturated salt water, one with dilute salt water, and one with distilled water. Invite predictions of how the manometer tubes will read on the following day.

Audio-Visual Aids

7-AV1. "Demonstrating the Gas Laws." Color. 21 min. Coronet. Boyle's, Charles', and the combined gas law.

7-AV2. "Gas Laws and Their Application." 13 min. Britannica. Early research leading to the description of the relationships between temperature, volume, and pressure of gases.

7-AV3. "Gas Pressure and Molecular Collisions." 20 min. Chem Study.

Relates gas pressure to absolute temperature and relative rates of effusion for various gases.

7-AV4. "Gases and How They Combine." Color. 22 min. Chem Study. Experimental evidence and a theory to support the Avogadro Hypothesis.

7-AV5. "Handling Gases." Color. 3 min. Yale University. Proper handling of high pressure commercial cylinders. Operating pressure reduction valves and releasing small quantities of gases from lecture cylinders.

7-AV6. "Hydrogen Fountain Demonstration." Color. 3 min. Yale University. Diffusion through clay filter.

7-AV7. "Kinetic-Molecular Theory." Color. 8 min. McGraw-Hill. Introduction to a theory for and the factors which influence the magnitude of kinetic energy.

7-AV8. "The Laws of Gases." Color. 11 min. Coronet. Properties of gases as described by Boyle, Charles, Dalton, and Avogadro. Determining the molecular weight of gases.

7-AV9. "Vapor Pressure." Color. Two titles: "Liquids," 4 min; "Solids," 3 min. Yale University. Demonstrating vapor pressure quantitatively. Vapor pressure as a function of temperature.

Ideas for Additional Instructional Activities

7-AD1. Alyea, Hubert M., "TOPS in Chemistry," *The Science Teacher,* 29: 14–21. September, 1962. Techniques, devices, methods, and experiments involving the properties of gases.

7-AD2. Bacon, W., "Determination of Vapor Density of Carbon Dioxide Assuming the Vapor Density of Air," *School Science Review,* 45: 421. 1963. Using two burettes and a clay pipe.

7-AD3. Berman, Robert H., and William N. Stakely, "Studying the Properties of Gases Experimentally," *Chemistry,* 38: 30–32. March, 1965. To show that the average kinetic energy of molecules of all gases is constant at a given temperature.

7-AD4. Chandler, E. E., "Rate of Diffusion," *Journal of Chemical Education,* 17: 513. 1940. Procedure adapts the HCl–NH_3 demonstration to allow Liesegang rings to form.

7-AD5. Coffin, C. C., "Experiments Simulating First and Second Order Gas Reactions," *Journal of Chemical Education,* 25: 167. 1948. Theoretical background relating first order reactions to the flow of a gas through an orfice and second order reactions to the flow of a gas through a capillary tube.

7-AD6. Cooper, M. J. R., "An Effusiometer for the Determination of Gas Densities," *School Science Review,* 43: 702. 1962. Apparatus for effusing gas through a capillary constructed from gas-washing tower. Includes equations, sample calculations, and representative data.

7-AD7. Davis, John D., "Study of Charles' Law," *The Science Teacher,* 26: 337–340. September, 1959. Operation of the law.

7-AD8. Davenport, D. A., "Hypodermic Syringes in Quantitative Elementary Chemistry Experiments: I, The Gas Laws," *Journal of Chemical Education,* 39: 252. 1962. Describes several types of syringes and needles and their use in demonstrating Graham's Law. Includes representative data.

7-AD9. Glasstone, Samuel, *Textbook of Physical Chemistry,* Princeton, New Jersey: D. Van Nostrand. Second Edition, 1946. Fundamental discussions of mean free path, collision diameter, collision with walls, effusion, and similar phenomena.

7-AD10. Hall, Marie Boas, "Robert Boyle," *Scientific American,* 217: 97–102. August, 1967. Features Boyle's approach to scientific proof.

7-AD11. Himes, D., "Diffusion," *Scientific Experiments in Chemistry,* New York: Holt, Rinehart, and Winston. 1959. Page 3. Reaction between hydrogen chloride and ammonia.

7-AD12. Hollahan, J. R., "Molecular Effusion: Its Newer Features and Applications," *Journal of Chemical Education,* 39: 23. 1962. Static effusion including use of diffusion to determine vapor pressure.

7-AD13. LeRoy, L. F., "A Quantitative Demonstration of Graham's Law of Diffusion," *Journal of Chemical Education,* 25. 215. 1948. Derives diffusion rate from pressure measured by the distance a water jet extends from a large bottle which is connected to a porous diffusion cup in an inverted bottle.

7-AD14. Reber, P., "Graham's Law of Diffusion Using an Overhead Projector," *Journal of Chemical Education,* 38: A23. January, 1961. Adapting the HCl-NH_3 reaction to overhead projection.

7-AD15. Wartburg, Arthur F., and James P. Lodge, Jr., "Estimating Concentration of Air Pollutants," *Chemistry,* 41: 29–31. February, 1968. Involves oxidants, sulfur dioxide, and suspended particle matter.

8

PERIODICITY IN THE PROPERTIES OF THE ELEMENTS

A chart displaying "The Periodic Table of the Elements" is very probably the most widely used audio-visual aid in the world's chemistry classrooms and laboratories. Beginning quite early in their study of chemistry, our students learn that this chart can be an ever increasingly useful source of information. The ability to fully exploit the facts, principles, and concepts which are incorporated in a periodic chart of the elements can be a clear-cut measure of many aspects of the success of our students.

The ever increasing precision with which the whole idea of periodicity in the properties of the elements has been described represents a classical episode in the history of chemistry. Our students, especially those who are interested in the social as well as the natural sciences, should be encouraged to trace the evolution of today's periodic table. Some of the instructional activities which follow are designed to achieve this goal.

INSTRUCTIONAL ACTIVITIES: STUDENT

8-1. Classification in Chemistry Versus in Biology. Organize a three-member debating team and prepare to defend or attack this statement: The classification system usually credited to Mendelyeev and based on the idea that certain properties of the elements are periodic functions of their atomic weights has turned out to be a more fruitful invention than has the system usually credited to Linnaeus in which the world's animals and plants are classified on the basis of similarities in their structures.

8-2. Periodicity in the Properties of the Elements as an Evolving Idea. Consult appropriate historical references and identify the specific observations which most likely led Mendelyeev to hypothesize "that there exists a kind of periodicity" in the properties of the elements. Include the information and ideas which Mendelyeev may have gained by reading the chemical literature of his day. Be prepared to debate the statement that Mendelyeev stumbled onto something actually much more significant than he realized. For example, it was not until 1913, forty-four years after Mendelyeev published his periodic table, that Moseley was able to say, "we have here a proof that there is in the atom a fundamental quantity, which increases by regular steps as we pass from one element to the next." And it was at least fifty years later that evidence was obtained to support the "magic number 8" which played such a vital role in Mendelyeev's creation—evidence indicating the high degree of stability that seems to accompany the existence of eight electrons in the outer domain of the atom.

8-3. Recognizing Periodicity in the Properties of the Elements. Limit your work to the elements with atomic numbers 1 through 18 and prepare a table listing the symbol of each element, its atomic number, its atomic mass, and the formulas of the hydrides, chlorides, and oxides of each element. What properties of each element seem to be repeated in the element 8 spaces farther up the atomic number scale? For example, what properties do elements 1 and 9 have in common? Elements 3 and 11? And so forth.

Beginning with elements having atomic numbers 3 and 4, prepare a similar table for elements 8, 16, 34, 52, and 84 spaces farther up the atomic number scale. For example, one portion of this table will involve elements numbers 3, 11, 19, 37, 55, and 87. Are there properties which all of these elements have in common?

The general periodicity in the properties of the elements follows the number series; 2, 8, 8, 18, 18, 32. To understand this concept, use a kind of "cumulative" thinking. For example, the element most similar to element number 1 is two spaces up the scale. For elements having atomic numbers greater than 2, their next "family" member will be 8 spaces ahead. Elements with atomic numbers above 10 but below 18 will find their next close relative 8 spaces ahead but elements with atomic numbers of 18 but not greater than 36 must move ahead 18 spaces to find another family member. Similarly, elements above atomic number 36 must jump 32 spaces to find another element with which it shares the maximum number of properties. Use this general expression to identify the elements in the same family as element number 2, element number 23, element number 13.

8-4. Trends in Properties Shared by a Family of Elements. This instructional activity involves four members of the alkaline earth family of elements. The activity encourages you to determine the heat involved in the reactions of the oxides of these elements with water and with 2.0 M hydrochloric acid. To help you interpret your data, handbook data tell us the relative volume of the ions of these elements, the tendency of each ion to attract electrons, and the amount of energy to strip the first electron away from atoms of each element.

Element	Ionic Radius (Å)	Electro-negativity	kcal/mole 1st Ionization Energy
Magnesium	0.65	1.2	176.2
Calcium	0.99	1.0	140.9
Strontium	1.12	1.0	131.2
Barium	1.34	0.9	120.1

Arrange with the other team at your lab station and prepare 200 ml of water in one calorimeter and 200 ml of 2.0 M hydrochloric acid in a second calorimeter. Obtain a vial containing one of the alkaline earth element oxides. Weigh the vial and then add approximately 4 grams of the oxide to the water calorimeter. While your partner is stirring the solution and recording the maximum temperature rise, reweigh the vial in order to determine the precise weight of the oxide you used. Now add carefully, a similar weight of the oxide to the acid in the second calorimeter, record the maximum temperature rise, and determine the precise weight of the oxide used. In all procedures involving the oxides of the alkaline earth oxides, keep to a minimum the exposure of the oxide to the moisture of the atmosphere.

Correct all of your calorimetric determinations for the calorimeter constants and scale your values up to kcal/mole of the oxide used. Add your data to the data table on the chalkboard. Examine the total data provided by the class and identify any orderly relationships among the data. Include the handbook data cited earlier in this activity.

INSTRUCTIONAL ACTIVITIES: TEACHER

8-T1. Periodicity in the Properties of the Elements Shown with Styrofoam Models. Students encounter the periodicity concept relatively early in the chemistry course, particularly as reflected in the valence of the elements. They ask for a theoretical explanation before an adequate background of atomic structure has been acquired. This dilemma calls for instructional activities which reveal those aspects of atomic theory which seem to be particularly associated with periodicity without necessarily going into the total topic of atomic structure.

A reasonably satisfactory instructional activity features the use of foam plastic balls to simulate the atomic orbitals. Single balls of various sizes simulate the *s* orbitals. Two balls with a rubber band drawn through their centers and secured at the ends with short lengths of pipe-cleaner can simulate a *p* orbital. Three of these can be entwined to simulate the complete *p* energy level. Similarly, two of these "orbitals" when entwined simulate hybridized *s* and *p* orbitals to produce the classical tetrahedron geometry.

These styrofoam models give the student something visible but it is still up to him to build his own mental models while we discuss the "filling" of the orbitals as we advance through the periodic table. The single *s* and the three *p* orbitals emphasize the eightness concept and "filled" versus "unfilled" or "partially filled" orbitals suggest positive versus negative valence.

By moving across rows of the periodic table and identifying the "address" of each new element's valence electrons, we bring out the periodicity concept. In effect, the same styrofoam structures repeatedly represent the electron architecture of the rows of elements. We merely change the names of the orbitals. The *d* and *f* orbitals, however, become so cumbersome that their simulation may be omitted relying on the other energy level simulations to have gotten across the repetition in electronic architecture essential to appreciation of the periodicity phenomenon.

8-T2. Trends in the Properties of the Members of the Alkali Metal Family of Elements. Looking ahead to reacting small masses of lithium, sodium, and potassium with water, place approximately 200 ml of water in each of three liter bottles. Add a drop of detergent and a few drops of phenolphthalein to each bottle. Experienced teachers working in well equipped laboratories may choose to add rubidium and cesium to this demonstration but the high degree of reactivity of these elements adds to the safety precautions which must be taken.

Protecting the bottles with adequate safety shields, add a mass of lithium the size of a split pea and immediately cover the bottle with a glass plate. Invite no more than two students to come as close as safety precautions permit and to share their observations with their classmates. Using the second bottle, repeat the procedure using an approximately equal mass of sodium. Continue with the potassium.

Anticipating that the first team of observers fail to catch all of the significant observations, e. g., the changing shape of the skittering mass of metal, have a second set of bottles at hand and do a re-run of the total presentation. And some of us chide the inadequate observers for their having allowed the obvious, grossly obtrusive observations to have kept them from seeing equally significant but more subtle observations.

After an adequate set of observations have been recorded, entertain explanations. Lead the students toward special attention to the rates of the reactions. If you are unduly attacked for the roughness in your estimates of equal masses of the three metals, be ready to demonstrate the volumes of 0.1 M hydrochloric acid needed to remove the indicator's pink color.

8-T3. Trends in the Properties of the Members of the Halogen Family of Elements. Prepare 0.1 M solutions of chloride, bromide, and iodide ions and demonstrate, without comment, that these solutions show no significant interaction with trichlorotrifluoroethane (TTE). Prepare solutions of chlorine, bromine, and iodine in water and again without comment, demonstrate the interactions of these solutions with TTE.

Announce that the next phase of this "silent demonstration" format will continue to provide background information but caution the students to avoid spoiling the eventual "punch line" by blurting out their hypotheses. Place 20 ml portions of each of the ion solutions in three sets of three large test tubes. To the first set, add 5 ml of trichlorotrifluoroethane to all three test tubes and then 20 ml of chlorine water to the first, 20 ml bromine water to the second, and 20 ml of iodine water to the third. Repeat this procedure for the second and third sets containing the bromide and iodide ion solutions.

See Figure 8-1 for a suggestion on how to display the results of each step in this demonstration effectively. Each step in the procedure is placed on the ascending spiral in order.

Continuing the "silent demonstration" format, prepare three sets of three test tubes in each set with the first test tube in each set containing equal volumes of the chloride and bromide solutions; the second test tube in each set containing equal volumes of the chloride and iodide solutions; and the third test tube with equal volumes of the bromide and iodide solutions. Prepare a fourth set containing equal volumes of all three ion solutions. Add 5 ml of TTE to all 12 test tubes.

Announce that you are about to add chlorine water to the first test tube in each set, bromine water to the second, and iodine water to the third but before you do so, each student is to sketch each of the twelve test tubes and predict the final condition, after shaking and settling, of each test tube. As soon as all predictions have been recorded, proceed with the additions of chlorine, bromine, and iodine water solutions to each of the designated test tubes.

Invite the students to reconcile their observations against their predictions but, continuing the "silent demonstration" format, to write an hypothesis that might be consistent with all of their observations and provide a satisfying description of the entire demonstration. Obviously, it is at this stage of the instructional activity that the "men and the boys begin to part company." Also, many students become increasingly anxious to verbalize their thoughts when they increasingly feel the bite of a puzzling situation. All of these behaviors combine to produce a pedagogical challenge which must be met if the instructional activity is to succeed and really give the students practice in processing their observations.

It is at this stage of the "silent demonstration" format that the teacher must also sense the kind and number of hints to use to "catalyze" student responses. Students who feel they were about to discover the answer resent suggestions that were too revealing. Students who have not yet caught even the slightest glimpse of what is going on resent being kept too long in a state of frustration. As they sometimes say, "You could have at least told us what TTE was doing in all of the test tubes." But there are few satisfactions

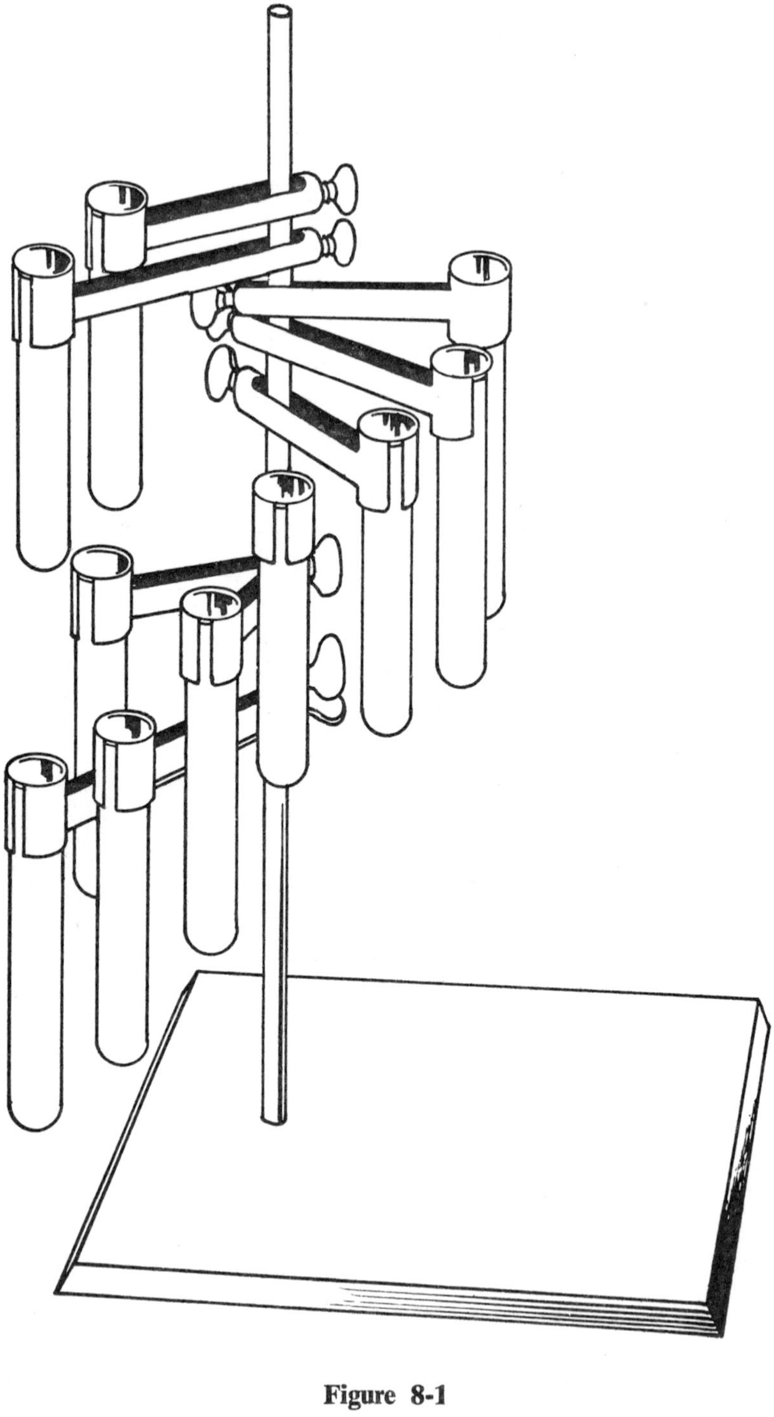

Figure 8-1

greater than seeing the look on those students' faces who "discover" that the members of the halogen family show decreasing ability to capture and hold electrons as we move down the periodic table.

Audio-Visual Aids

8-AV1. "Atomic Models, Valence, and the Periodic Table." Color. 44 min. State University of Iowa. Elementary treatment of electronic architecture, polarity, and related topics.

8-AV2. "Chemical Families." Color. 21 min. CHEMS. Featuring the halogen and alkali metal families, shows experimentally obtained evidence of similarities in properties within families.

8-AV3. "Ionization Energy." Color. 22 min. CHEMS. Experimentally derived evidence of ionization energies and relationship between ionization energy and chemical reactivity.

8-AV4. "Ionization Potential and Electronegativity." Filmstrip of the Month Club. 355 Lexington Ave., New York, New York 10017. Meaning of ionization potential, electron affinity, and electronegativity.

8-AV5. "Metals and Non-Metals." Color. 9 min. Yale University. In relation to placement in the periodic table.

8-AV6. "Relative Size of Atoms." Chart. Welch. 42 × 58 inches. Portrays atomic and ionic volumes with radii shown quantitatively.

8-AV7. "Relative Sizes of Ions." 32 frames. Color. Britannica. Ionic size portrayed by packings of spheres.

8-AV8. "Welch Chart of the Atoms." 42 × 58 inches. Color. Welch. One of the standard versions of the periodic table. Includes a maximum amount of information in minimum space.

Ideas for Additional Instructional Activities

8-AD1. Donnelly, L. R., "Reactions Between Oxides and Water," In *Scientific Experiments in Chemistry,* New York: Holt, Rinehart, and Winston. 1959. Page 5. Leading to classification as acid- or base-forming.

8-AD2. Dutton, F. B., "Oxidation of Bromide and Iodide Ions," *Journal of Chemical Education,* 40: A241. March, 1963. Using chlorine water with potassium iodide and bromide.

8-AD3. Feigl, F., "Spot-Reaction Experiments: V, Solvent Effects," *Journal of Chemical Education,* 20: 300. 1943. Solutions of iodine in carbon disulfide and in ether are reacted with strips of copper, with silver cyanate, and with mercury(II) chloride.

8-AD4. Furst, A., "Displacement in the Halogen Series," *Journal of Chemical Education,* 24: 445. 1947. Action of bromine water and chlorine water on solutions of sodium hypochlorite, sodium hypobromite, and sodium iodide. Includes preparation of the halogens.

8-AD5. Jones, H. R., "The Classification of Oxides," *School Science Review,* 44: 95. 1962. Decomposition of the carbonates of magnesium, strontium, and barium in terms of delta-H and delta-S, lattice energies, delta-G, and radii of cations and anions.

8-AD6. Mahan, B. H., "A Simple Ice Calorimeter," *Journal of Chemical Education,* 37: 634. 1960. Construction of a calorimeter than can be adapted to comparing enthalpy of reactions of alkaline earth oxides with hydrochloric acid.

8-AD7. Masterton, W. L., and J. J. Demo, "Reduction of Oxides By Hydrogen," *Journal of Chemical Education,* 35: 242. May, 1958. Suggests a method for quantitative comparison of the reduction of such oxides as copper(II), nickel(II), cobalt(II), tin(IV), and copper(I).

8-AD8. Pauling, Linus, *The Nature of the Chemical Bond,* Ithaca, New York: Cornell University Press. Third Edition, 1960. Pages 88–97. Relating electronegativity to other properties.

8-AD9. Rossini, F. D., *et al,* "Selected Values of Chemical Thermodynamic Properties," in *National Bureau of Standards Circular 500,* Part I, United States Government Printing Office, 1961. Enthalpy of formation, free energy of formation, entropy, heat capacity, and related properties.

8-AD10. Sanderson, R. T., *Chemical Periodicity,* New York: Reinhold. 1960. Pages 116–117. A scale of electronegativities labeled as stability ratios.

8-AD11. Seeger, W., "Simple Determination of the Equivalent Weight of Mercury, Copper, and Zinc," *Journal of Chemical Education,* 16: 595. 1939. Treating known masses of the metallic oxides with excess standard hydrochloric acid and titration of the excess with standard base.

8-AD12. Weaver, Elbert C., "Heating Metallic Oxides," In *Scientific Experiments in Chemistry,* New York: Holt, Rinehart, and Winston. 1959. Page 55. Compares tendency of metallic oxides to decompose when heated.

9

SOLUTIONS AND SOLUBILITIES

Solutions and solubility phenomena enter into the everyday experiences of students in so many ways that they, for sure, have common sense knowledge of the properties of solutions and the factors which affect the solubilities of substances. Everyday experiences with solutions, however, seem to fail to generate the realization that the solution process in itself may be a major step in the ultimate reshuffling of atoms in a chemical reaction.

Instructional activities are called for which show that molecules and ion pairs can be not only separated but actually fragmented during the solution process. It follows that the separate ions or fragments may have properties quite different from those of their undissolved sources. And not always are these original properties restored when the solvent is removed and the dissolved substances restored to the solute status.

Curiosity is easily aroused by solution phenomena. Students wonder why substances are or are not soluble in a given solvent. Differing degrees of solubility also arouse curiosity. Questions such as these invite exploration of the interactions between particles of solvent and solute—interactions likely to involve the relative polarities of all particles. In systems containing more than one solvent or more than one solute, these interactions can become quite complex. It is necessary to consider some of these complex systems, however, if the students are to grasp the nature of solution processes.

The properties of solutions which stem primarily from the relative concentration rather than the kinds of dissolved particles, the colligative properties, arouse the interest of students. In addition to being useful for estimating the molecular weights of solutes, these properties also can be used to focus attention on the nature of solutions and the solution process. So do the heat effects accompanying the formation of solutions although these effects are considered in greater detail in Chapter 19.

INSTRUCTIONAL ACTIVITIES: STUDENT

9-1. Handbook Information Dealing with Solutions and Solubilities. To put the reacting substances into solution is the first step in bringing off a great many chemical reactions. On this basis, handbooks which provide the kind of information most likely to be useful to chemists often include the weight in grams of selected substances which can be dissolved in specified volumes of selected solvents, especially water. The following table shows a sample of this kind of information. In this table, the numbers appearing as superscripts indicate the actual temperatures rather than the more general notation of *cold* versus *hot* water.

Name	Formula	Formula Weight	Solubility, in grams per 100ml Cold Water	Hot Water
Aluminum sulfate	$Al_2(SO_4)_3$	342.15	31.3^{0}	98.1^{100}
Barium sulfate	$BaSO_4$	233.40	$2.2 \times 10^{-5\,18}$	$3.3 \times 10^{-5\,50}$
Calcium acetate	$Ca(C_2H_3O_2)_2$	158.17	37.4^{0}	29.7^{100}
CeriumIIIsulfate	$Ce_2(SO_4)_3$	568.42	10.1^{0}	2.25^{100}
Lead acetate	$Pb(C_2H_3O_2)_2$	325.28	44.3^{20}	221^{50}
Lithium hydroxide	$LiOH$	23.95	12.8^{20}	17.5^{100}
Magnesium hydroxide	$Mg(OH)_2$	58.33	$9 \times 10^{-4\,18}$	$4 \times 10^{-3\,100}$
Potassium hydroxide	KOH	56.11	107^{15}	178^{100}
Sodium hydroxide	$NaOH$	40.00	42^{0}	347^{100}

Are there orderly relationships within the data shown by this table? That is, are there any statements you can make about the solubilities of substances that are not contradicted by the data for at least one of the substances in the list? Don't be surprised if you come to the tentative conclusion that the solubilities of substances are pretty much random properties. But it would be surprising to find that solutions and solubility phenomena are entirely unexplainable in the sense of not fitting into any system of scientific principles or laws.

Let's play with two rather crude ideas or models whereby we might "explain" solution phenomena. First, let's think of a solute going into solution something like golf balls or marbles "dissolving" in a container filled with volley balls. In effect, this model explains the solution process as allowing the solute particles to occupy spaces between the solvent particles. Let's call this our "niche filling" model.

For a second model, imagine the solvent particles as having some kind of attractional force which causes the solute particles to become attached and thus held distributed throughout the solvent. Let's call this our "sticky" model. Are there bits of information about the solubilities of substances as shown in the table that are better "explained" by one than the other of these two models? Are there facts for which neither model provides a logical "explanation"?

Which model best "explains" the effects of temperature changes on the solubilities of substances? Remember that temperature provides a measure of molecular motion.

What does it do to your thinking to be reminded that 100 ml represents approximately 5.5 moles of water particles each of which has a tendency to "stick" to each of its neighbors? For example, is there always a definite ratio between the number of solvent particles and the number of solute particles held in a saturated solution? Or is it true that the more massive the solute particles, the greater the number of solvent particles which must "team up" to keep one solute particle in solution? When you are thinking through this question, remember that many ions do not break up when they go into solution. Sulfate, acetate, and hydroxide ions are likely to remain as particles.

9-2. Electrical Conductivity as a Clue to the Nature of Solutions and Solubilities. Place 20 ml of distilled water in one and 20 ml of ethanol in another 50 ml beaker. Use the conductivity tester to determine whether or not these two solvents conduct the electrical current. Go to the checkout table and obtain one of the one gram samples of the substances at the table. Divide this sample equally and add 0.5 gram to each of your two solvents, the water and the alcohol. Stir thoroughly and record your estimate of the amount of the solid material that dissolves.

Use the conductivity tester to determine whether or not your two "solutions" conduct the electrical current. Be sure to rinse the electrodes both before and after testing each solution. Enter your data on the data table on the chalkboard. Copy the data obtained by all of your classmates. A representative data table might be:

Substance	Water How much dissolved	Conductivity	Alcohol How much dissolved
Paradichlorobenzene	None	None	Very little
Fumaric acid	Very little	None	Some
Benzoic acid	Very little	None	All
Aluminum sulfate	All	Good	Very little
Sodium hydroxide	All	Good	Very little

For a portion of this instructional activity, your instructor will show you whether or not the substances from which you have made solutions conduct electricity. Add this information to the data table and use all of the data to predict the degree to which each of the following statements does or does not apply to solutions and the solubilities of substances. (1) If a substance is soluble in a solvent, the resulting solution conducts the electrical current. (2) A solvent that does in itself conduct electricity may lose the conductive property when a substance is dissolved in it. (3) A solvent that does not in itself conduct electricity may produce a conducting solution when any kind of solute is dissolved in it. (4) No substance which, when in the pure, solid form does not conduct electricity, produces a solution which does conduct the electrical current. (5) It is highly probable that the dissolving of a substance involves the electrical balance between the particles of the yet to be dissolved substance.

9-3. Solutions of Substances Versus Substances in the Liquid State. Use the conductivity tester to determine whether or not each of the following substances conducts electricity when (1) in the pure, solid form, (2) dissolved in water, and (3) when heated to the melting temperature: paradichloribenzene, diphenylamine, potassium nitrate, sodium acetate, fumaric acid, maleic acid, potassium chromium sulfate.

Are there substances which form conducting solutions but do not conduct electricity when in the liquid state? Are there substances which conduct electricity when melted but, when dissolved, their solutions are non-conducting? Thinking through the evidence from instructional activities 9-1, 9-2, and 9-3, is there reason to believe that the solution process is more likely to involve bonds between molecules than the bonds which hold atoms together within molecules? To understand the solution of any substance and its relative solubility, would it be helpful to know whether the bonds between its particles are more ionic than covalent or vice versa?

9-4. Solution: An Energy Involving Process. Using a styrofoam cup or a more efficient calorimeter, determine the temperature change when 6.02 grams of magnesium sulfate are dissolved in 100 ml of water. Similarly, determine the temperature change when 12.32 grams of magnesium sulfate seven hydrate are dissolved in 100 ml of water. When doing your calculations, estimate the heat capacity of the magnesium sulfate solutions to be 0.92 cal/gram. See if you can devise a way to correct your data to account for the heat lost or gained by a calorimeter during its use for a calorimetric measurement.

Scale up the values you have obtained so you can estimate the heat involved per mole of anhydrous and hydrous magnesium sulfate going into solution. With what factors might we relate this heat involvement? For example, would energy be absorbed or released in breaking down the crystal structure of the magnesium and sulfate ions preliminary to going into solution? If so, which of the two forms of the substance releases or absorbs the greater amount of energy in this step? Assuming that the ions of the substance interact with water molecules upon forming a solution, is it possible to allocate the proper fraction of the net heat involvement to the two steps; the breaking down of crystal structure and the interaction of the ions with the molecules of the solvent? If not, what argument would infer that the breaking down of crystal structures is the more significant?

Looking back through this instructional activity, is there evidence to support the statement that the solution process not only may involve the breaking of bonds between and within molecules or crystal systems but also the formation of new bonds or some other kind of interaction with the particles of the solvent? Is it a valid assumption that changes in heat content are evidence of bond breakings and reformings?

Again, using a styrofoam cup or other calorimeter, determine the heat content changes that accompany the solution of 5.35 grams of ammonium chloride in 100 ml of water. Scale up your value to kilocalories per mole. Contrast the involvement of heat in the solution of ammonium chloride with that of magnesium sulfate. For example, is there evidence of breakdown of crystal structure in both cases? If we assume that energy is absorbed during this step of the solution process, what must we conclude must be the most striking difference between the heat changes accompanying the solution of these two salts?

9-5. Interactions Between Solutions. The burettes at the check-out table contain 0.5 M solutions of either lead nitrate or sodium iodide. On the basis of the indicated concentration of these solutions, there should be 0.165 gram of lead nitrate per ml of the first solution and 0.075 gram of sodium iodide per ml of the second solution. Label two clean, dry, and pre-weighed evaporating dishes (pencil on bottom) and transfer 5 ml of the solutions to separate evaporating dishes. Verify the weights of solutes in the samples of the two solutions by evaporating the solvent. To speed up evaporation, either place the dishes on thick asbestos mats on heaters or use the dishes as "lids" on beakers in which water is boiled until evaporation of the solutions is achieved.

While the above samples of the two solutions are evaporating, transfer 10 ml samples of each solution to two 100 ml beakers. Combine the two solutions. Stir briefly. Using two thicknesses of filter paper, filter off the precipitated yellow lead iodide. Use only as little distilled water as necessary to wash the residue from the beaker onto the filter paper.

While the filtration process is going on, calculate the weight of lead iodide that should be present on the filter paper. (Teachers differ in their willingness to paternalistically remind their students to retrieve prior information, e. g., the formula of lead iodide is PbI_2, rather than allowing the forgetful student to suffer the effects of not using his wits to avoid goofs.) Anticipating the possibility that the actual yield of lead iodide turns out to be less than the calculated yield, save the filtrate so you can investigate the apparent discrepancy between actual and calculated yields of lead iodide.

Assuming that you cannot complete this activity during a single lab period, plan to allow the filter paper containing the lead iodide to dry over night. Similarly, allow the filtrate to evaporate as much as it will over night. On the following day, reexamine all of your data and state a problem summarizing any discrepancies which may exist between actual and calculated yields of lead iodide. Be ready to state various hypotheses which might account for the discrepancies.

One such hypothesis may suggest that whereas lead nitrate and sodium iodide are readily soluble in water, lead iodide is only sparingly soluble in water. Think about the logic reflected in this hypothesis. It infers that water molecules can, apparently, break down the crystal systems of lead nitrate and sodium iodide but they cannot completely prevent new crystal systems of lead iodide from forming. Furthermore, since we are assuming that the precipitate contains no sodium nitrate, our hypothesis infers that water particles can prevent the formation of new sodium nitrate crystal structures.

How might we investigate the validity of this hypothesis? If you wish to exercise your own ingenuity in designing experiments, do not read ahead until you have attempted to answer this question on your own. One approach begins by consulting handbooks to see if lead forms compounds which are less soluble than is lead iodide. We may find, for example, that whereas 100 ml of water at 20°C can dissolve 0.063 gram of lead iodide, the same volume of water can dissolve less than 6×10^{-6} gram of lead chromate at the same temperature. On this basis, when the lead iodide was filtered, the 20 ml or so of filtrate could contain 0.012 gram of dissolved lead iodide but nearly all of the lead ions would be precipitated as lead chromate if chromate ions are added to the solution. Verify this prediction by adding, drop by drop, 0.1 M potassium chromate to the filtrate. The predicted yield, 0.01 gram, of lead chromate, however, may be undetectable unless you have balances sensitive to milligrams.

9-6. Additional Interactions Between Solutions. Obtain 20 ml portions of 0.1 M hydrochloric acid and 0.1 M sodium bicarbonate. Combine the two solutions. Assuming the bubbles which appear are carbon dioxide and that the sodium bicarbonate solution contained bicarbonate ions, what can we infer regarding the solubilities of the various particles which occur initially and finally when the acid and soda solutions were combined?

Place approximately 5 ml of 6 M hydrochloric acid in a gas collecting tube. Carefully fill the tube with water avoiding in so far as possible mixing the acid with the water. Cork the tube and invert. Consult a dictionary for the meaning of *schlieren* and if this phenomenon is observed, what evidence regarding the nature of the solution process does this activity provide?

To 15 ml of 0.002 M potassium thiocyanate, add a drop at a time 0.2 M ferric nitrate. How does the interaction between these solutions differ from and how is it similar to the interaction between lead nitrate and sodium iodide? To account for the color changes when the potassium thiocyanate and ferric nitrate solution are combined, hypothesize the formation of thiocyanoiron(III) ions, $FeSCN^{+2}$. If these ions exist in the combined solutions, does it follow that the solvent water particles cannot keep the ferric and thiocyanate ions apart even though the resulting combination retains unbalanced electrical charge and remains in solution?

To 15 ml of 0.1 M copper sulfate add a drop at a time concentrated ammonia-water solution. See if you can put together a complete explanation of all of your observations if you know that (1) copper ions in water tend to be surrounded by 4 water molecules, (2) the hydroxides of many metals are insoluble or, at best, have low solubilities, and (3) ammonia molecules have many of the polar characteristics of water molecules.

INSTRUCTIONAL ACTIVITIES: TEACHER

Some teachers will prefer to present many of the instructional activities listed for students as demonstrations. Similarly, the following teacher activities may be adapted for laboratory exercises or individual student projects. In general, activities which are inherently time consuming or are likely to involve unanticipated safety precautions seem to justify being kept pretty much under the control of the teacher. In addition, teachers learn to anticipate the features which need special emphasis and can highlight these features in the demonstration format.

9-T1. Calculations Involving Solubility Products. Assuming your students have been introduced to the qualitative expression for solubility products as a general concept, present the expression for the solubility product of silver chloride on the chalk board. Also, indicate its numerical value. Without further comment, place 1 liter of tap water in a suitable container and begin adding silver nitrate solution drop by drop from a container clearly labelled 0.001 M $AgNO_3$. Count the drops rather dramatically and stir the system thoroughly between drops. Be sure the system is illuminated in such a way that the students can discern clearly when the faint haze of a white precipitate remains after stirring. Announce that each drop of the silver nitrate solution represents approximately

0.055 ml and challenge the students to determine the concentration of chloride ion in the tap water.

If the response to the challenge is delayed unduly, remind the students of the dilution formula:

$$\text{Volume}_{(\text{initial})} \times \text{Molarity}_{(\text{initial})} = \text{Volume}_{(\text{final})} \times \text{Molarity}_{(\text{final})}$$

A similar challenge may be set by such a problem as: Suppose that the sulfate ion concentration in a stream is approximately 1×10^{-6}M and that the rate of flow is 1×10^{6} liters per minute. The stream flows past a factory that plans to dispose of wastes containing calcium ion. What is the maximum weight of the calcium ion that can be dumped per minute into the stream without adding to the turbidity of the stream? Assume the initial concentration of calcium ion in the stream to be zero and be ready to discuss additional assumptions which must be taken into account to make this a logically sound problem.

Many students have the rather healthy tendency to try to take apart problems which show an effort to deliver a message along with providing practice in applying fundamental computational skills. This tendency was anticipated in the problem discussed in the previous paragraph. When the message, in this case a message involving stream pollution, is a legitimate extension of the course content, the discussion of the logic of the problem is really intended to emphasize the application of knowledge toward the solution of significant social or economic problems.

9-T2. The Colligative Properties of Solutions. Have on hand several liters of a freezing mixture consisting of three parts crushed ice and one part rock salt. When the class convenes, promptly prepare 100 ml volumes of 2.0 F solutions of sodium chloride, calcium chloride, methanol, and sugar. Assign each pair of students to determine the freezing or boiling point of a 10 ml sample of one of the solutions. As rapidly as the data are obtained, enter each value in a chalkboard data table. Challenge the students to "discover" orderly relationships within the total data table. Hopefully and keeping in mind the confidence limits within which the thermometers being used can be read, the students will find that the boiling point is elevated approximately 0.51C° and the freezing point is depressed 1.86C° per mole of dissolved particle in one liter of water.

The more capable members of the class may be given the additional assignment of seeking orderly relationships in data obtained from determining the freezing points of systems consisting of various weights of sulfur dissolved in naphthalene. Handbooks give the molecular depression of the freezing point of naphthalene to be 68–69C° for one gram molecular weight dissolved in 100 grams.

9-T3. The Solution of Hydrogen Chloride in Water. Generate hydrogen chloride by the usual method and allow the gas to dissolve in water by placing a funnel on the delivery tube from the generator with the mouth of the funnel a few millimeters above the level of the water. Advise the students to observe the schlieren and then invite a discussion of the causes of this phenomenon. Use the discussion to emphasize the factors which influence the solubilities of gases in liquids (Figure 9-1).

Before dismantling the set-up, collect a 500 ml flask of the hydrogen chloride gas and use an adaptation of the "ammonia fountain" to demonstrate not only the high solubility of this gas in water but also to drive home the safety precaution involved in removing delivery tubes from pneumatic troughs when generating water soluble gases in heated glass generators. To achieve the fountain effect, fit the flask with a two-hole stopper carrying a jet and a water-filled medicine dropper. See Figure 9-2.

Audio-Visual Aids

9-AV1. "Ammonia Fountain Demonstration." Color. 2 min. Yale University. Reduced atmospheric pressure within a flask resulting from the high solubility of ammonia in water.

9-AV2. "The Colloidal State." Color. 16 min. Coronet. Nature, preparation, and uses of.

9-AV3. "Colloids." 11 min. Britannica. Types of, factors influencing colloidal systems, the Tyndall effect, Brownian movement, catephoresis, the Cottrell process.

9-AV4. "Electric Interactions in Chemistry." 21 min. Chem Study. Experimentally derived evidence of the relationships between electric interactions and distances separating charged objects.

9-AV5. "Heat of Solution." Color. 16 min. Coronet. Determination of calorimeter constant. Calculation of heat of solution of zinc sulfate.

9-AV6. "Ionization." Color. 18 min. Coronet. Theory and applications of by laboratory demonstrations and animation.

9-AV7. "Ionization and Electrolysis." 51 frames. Long Filmstrip Service.

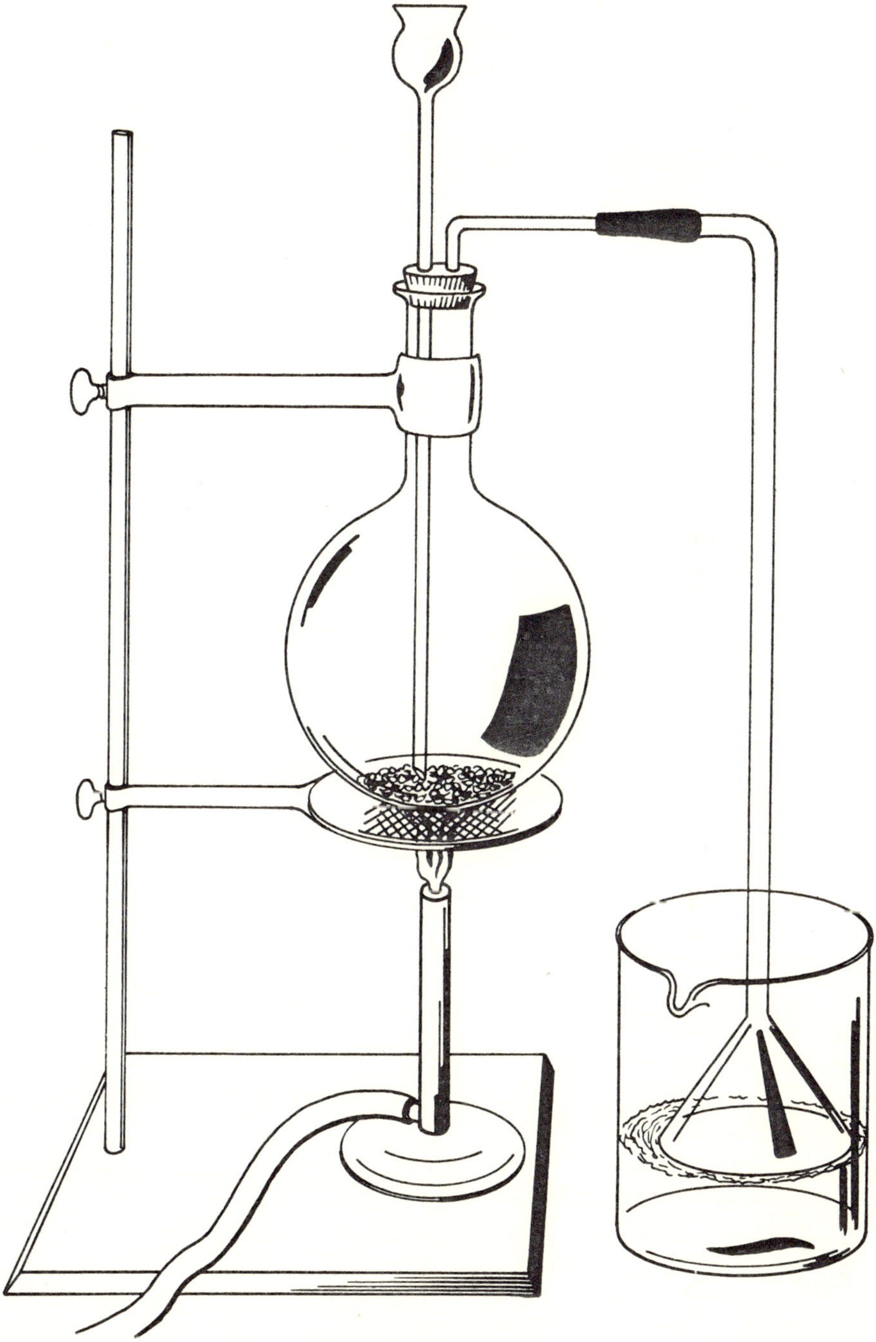

Figure 9-1

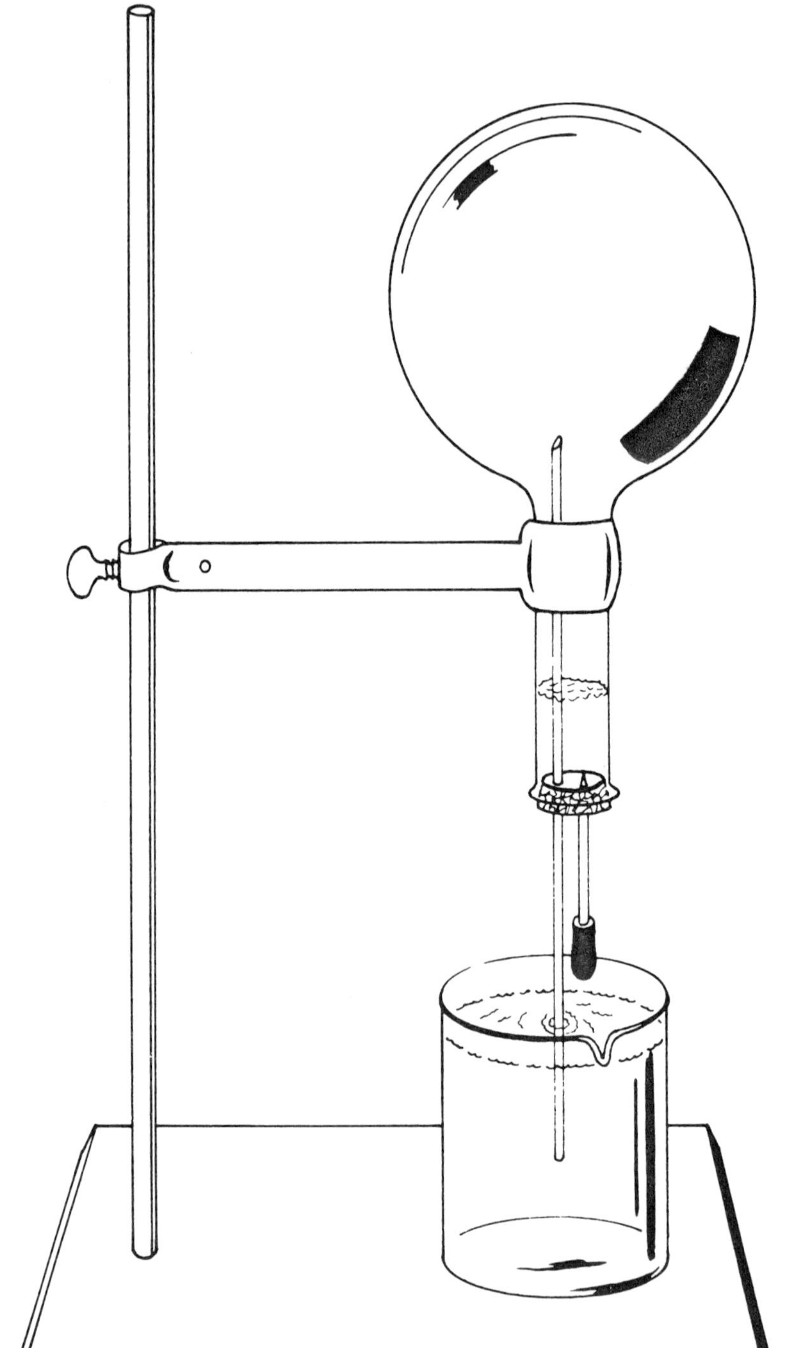

Figure 9-2

Ions versus atoms and the roles ions play in several types of reactions.

9-AV8. "Molecular Weight of Solutes." Color. 8 min. Coronet. Based on colligative properties.

9-AV9. "Principles of Ionization." Color. 13 min. Coronet. Effects of solutes, ion migration during electrolysis.

9-AV10. "The Solution Process." 40 frames. FOM. Interactions between solute and solvent and accompanying energy changes during the dissolving process.

Ideas for Additional Instructional Activities

9-AD1. Alyea, Hubert N., "Molecular Weight from Freezing Point Lowering," *Journal of Chemical Education,* 43: A349. April, 1966. See also, "Vapor Pressure and Raoult's Law," 37: 57. 1960.

9-AD2. Alyea, Hubert N., "Crystallization Techniques," *Journal of Chemical Education,* 43: A135–A136. February, 1966. Benzoic acid, cupric sulfate, mercuric chloride, oxalic acid, potash alum, potassium chlorate, potassium nitrate, sodium chloride with and without urea, potassium acid phthalate, Rochelle salt, and sucrose. Also, mending damaged crystals.

9-AD3. Andrew, H. G., "Negative Temperature Coefficient of Solubility," *School Science Review,* 45: 405. 1963. Preparation of zinc phosphate and investigation of its enthalpy of solution.

9-AD4. Bacon, E. K., "Use of Calcium Acetate in the Demonstration of a Substance Showing a Negative Coefficient of Solubility," *Journal of Chemical Education,* 15: 494. 1938. Investigates the drop in solubility from 52 g/100 g of water at 0°C. to 29.7 g/100 g of water near the boiling point.

9-AD5. Barrett, Richard L., "Solubility Product Constant for Cadmium Hydroxide," *Journal of Chemical Education,* 36: A501–503. 1959.

9-AD6. Bishop, J. A., "The Solubility Product: A Proportionality Constant," *Journal of Chemical Education,* 31: 574. 1954. Graphic approach to solubilities of silver chloride, lead(II) chloride, antimony(V) sulfide, and zinc sulfide.

9-AD7. Davis, J. C., "Acetamide as a Solvent for Freezing Point Depression and Solubility Experiments," *Journal of Chemical Education,* 43: 611. November, 1966.

9-AD8. Eddy, R. D., "The Solubility Curve of Borax," *Journal of Chemical Education,* 35: 364. July, 1958.

9-AD9. Haggin, Joseph H. S., "The Candy Chemist," *Chemistry,* 40: 8–12. April, 1967. Controlled crystallization.

9-AD10. Hanson, A. L., "Physical Properties of Liquid Mixtures," *The Science Teacher,* 28: 43. 1961, February. Representative data involving surface tension and viscosity.

9-AD11. Hoare, J. P., "Freezing Point Measurement," *Journal of Chemical Education,* 37: 146–147. March, 1960. Representative data for chromium(IV) oxide in sulfuric acid.

9-AD12. Jones, Jerry L., "Precipitation from Mixed Solvents," *Journal of Chemical Education,* 45: 433–434. June, 1966. Separation by controlled precipitation.

9-AD13. Kauffman, George B., "Alfred Werner, Architect of Coordination Chemistry," *Chemistry,* 39: 14–18. December, 1966.

9-AD14. Klemm, L. H., "Some Lecture Demonstrations in General Chemistry," *Journal of Chemical Education,* 28: 587. November, 1951. Uses lead(II) nitrate and sodium bromide to investigate formation of lead(II) bromide.

9-AD15. Kobe, K. A., "Demonstration of an Inverted Solubility Curve," *Journal of Chemical Education,* 16: 183. 1939. Investigates the formation of various crystal hydrates by saturated manganese(II) sulfate solution containing 0.5% sulfuric acid when the solution is cooled and/or heated. For example, a solution saturated at 27°C. yields pink crystals of manganese(II) sulfate seven hydrate in an ice cooling mixture and white crystals of manganese(II) sulfate one hydrate at 100°C.

9-AD16. Kokes, R. J., M. K. Dorfman, and T. Mathia, "Experiments for General Chemistry. V. Equilibria in Ionic Solutions," *Journal of Chemical Education,* 39: 93. February, 1962. Representative data for solubility product of silver chloracetate.

9-AD17. Lange, N. A., *Handbook of Chemistry.* New York: McGraw-Hill. 1961. 10th Edition, page 1502. Among other valuable data, heats of solution for various compounds.

9-AD18. Lee, J. H., "Hydrophilics," *School Science Review,* 43: 391. 1962. Uses the continuous variation type of experimental design to investigate the formation of calcium hydroxide from calcium chloride and sodium hydroxide.

9-AD19. McClenon, Robert, "Chemistry in Silica Gel," *Chemistry,* 40: 30–32. February, 1967. Diffusion and crystal growth.

9-AD20. Miller, J. G., and W. Lucasse, "Freezing Point Experiments for the Undergraduate," *Journal of Chemical Education,* 17: 522. 1940. Sucrose for molecular mass determination, resorcinol for association phenomena, potassium chloride and magnesium chloride seven hydrate for electrolytic solutions. Supercooling.

9-AD21. Nechamkin, Howard, "Solubility Product," *The Science Teacher,* 29: 45. March, 1962. Based on formation of calcium chromate by reaction of chromate(IV) ion with iodide ion and titration of the free iodine with sodium thiosulfate.

9-AD22. Palit, S. R., "Solvent Action and 'Co-Solvency'," *Journal of Chemical Education,* 23: 182. 1946. Sodium oleate does not dissolve in either propylene glycol or chloroform but does dissolve in a mixture of these two solvents.

9-AD23. Palmer, W. G., *Experimental Physical Chemistry.* New York: Cambridge University Press. 1954. Contains many demonstrations suitable for investigating solubility phenomena. For example, using ammonium oxalate for solubility curve, sodium carbonate ten hydrate to show transition to other forms, and potassium iodide containing mercury(II) iodide for freezing points. Preparation of freezing baths.

9-AD24. Peterson, B. H., "The Solubility Product Constant for Copper Iodate: An Experiment," *Journal of Chemical Education,* 34: 612. 1957. Titrimetric analysis with sample calculations.

9-AD25. Snyder, Asa E., "Desalting Water by Freezing," *Scientific American,* 207: 41–47. December, 1962. Freezing crystals of fresh water from salt water.

9-AD26. Stong, C. L., "Growing Crystals in Silica Gel Mimics Natural Mineralization," *Scientific American,* 206: 155. March, 1962.

9-AD27. Wolthuis, Enno, M. Visser, and J. Oppenhuizen, "Molecular Weight Determination by Boiling Point Elevation," *Journal of Chemical Education,* 35: 412. August, 1958. Apparatus and procedure.

10

ACIDS AND BASES

When students enter the chemistry course, their knowledge of acids and bases may be characterized by such expressions as: Acids "eat holes in things." Bases, especially strong alkalis, clean clogged drains. It is safe to assume, usually, that these students have not yet acquired an appreciation of any theoretical basis for acid-base interactions and that giant steps must be negotiated before "proton donor" or "Bronsted-Lowry" are meaningful terms.

Interactions between electrons and protons are central to the description of almost any chemical event. More or less unattached protons are effective "threats" to any existent electron-proton bond. It follows that any substance which frees protons can play an active role in rearranging atoms within molecules. Similarly, any substance that provides forces which tend to dislodge protons also threatens to break existent bonds within molecules. It follows that acids and bases are quite likely to "eat holes in things" or otherwise convert insoluble, stable arrangements of atoms in molecules to new soluble or otherwise less stable particles. In effect, acids and bases can threaten the "glue" which holds the world of matter together.

Many students are impressed by the apparent paradox that water, plain, everyday, life-sustaining water, combines the two particles which play the leading roles in the actions of acids and bases. Exceedingly significant phenomena derive from changes in the hydrogen ion concentration in water. On this basis, it is especially important that our instructional activities improve student comprehension and appreciation of the nature and action of acids and bases.

INSTRUCTIONAL ACTIVITIES: STUDENT

10-1. "Anything Water Can Do, Acids and/or Bases Can Do Better." Can you translate the somewhat flippant title of this instructional activity into a more formal problem statement? Keep in mind that water dissolves many different kinds of substances, it provides a medium in which reactions can occur, and it reacts with certain substances.

You will find a variety of substances and several different kinds of acids and bases on the check-out table. Survey the materials that are available and plan experiments whereby you can test the hypothesis that is suggested by the title of this instructional activity. For example, what happens to the speed of a reaction between water and another substance when either a 6 N acid or base is substituted for the water? If a substance dissolves in water, in an acid, or in a base, is the same residue left after evaporation of the liquid portion of the system?

Note: Provide both strong and weak acids in 6 N and 1 N concentrations in dropper bottles (*safety glasses*). Baking soda; various kinds of metals, especially calcium; various indicators; textile samples; and other organic and inorganic substances may be provided.

10-2. Getting Acquainted with Acids. Examine the labels on the containers in which the acids used in the laboratory are delivered. Although you may be interested in the other information on the label, copy only the information that deals with the concentration of the contents. Calculate the weight of 1 liter of all of the acids which are liquids and then determine the number of moles of the labelled substance in one liter. If the acids are solids, calculate the weight needed to prepare 1 liter of each acid having 1 M concentration.

Examine the formulas of all of the acids. In how many cases can you "subtract" one or more water molecules from the formula? In some cases, try doubling or tripling the formula. In general, what kind or class of compound remains after subtracting one or more water molecules from the acid's formula? Looking again at the formulas of the acids, how many hydrogen atoms do you find in each acid formula? Calculate the volume of each liquid acid that would be needed to provide one mole of hydrogen atoms. What should be the concentration of hydrogen ion for each of the acids if these calculated volumes from the stock bottles were added to enough water to make one liter total volume? What assumptions are involved in your predictions? Have your teacher discuss the validity of these assumptions.

10-3. Getting Acquainted with Bases. Adapt Instructional Activity 10-2 and use the general approach to study the bases that are available in your laboratory. Realize that whereas the action of acids depends very much on hydrogen ions, it is the hydroxide ions that play the central role in the action of bases.

Suppose you were to find substances with such formulas as $Al(OH)_3$ or $Sn(OH)_4$. How would these substances differ from substances having the formulas H_3AlO_3 and H_4SnO_4? Can you think of any reason why the same substance might produce hydrogen ions under one set of conditions and hydroxide ions under a different set of conditions? From a metaphorical point of view, can you think of any situation in which a rich man might receive a handout from another rich man? Or a poor man might give money to another poor man?

10-4. The Interaction of Acids and Bases. When acids and bases interact, the hydrogen ions from the acid can combine with the hydroxide ions from the base to form water. It is equally logical to assume that the positive ions from the base will combine with the negative ions from the acid to form a salt. But some salts are soluble and will remain as ions in solution. Only when the salt from the neutralization of an acid by a base is insoluble does the appearance of a precipitate accompany the interaction between acids and bases. Sometimes this kind of interaction produces a gaseous product which, after saturating the solution, bubbles escape from the system.

Practice writing the equations for reactions involving all of the acids and bases available in your laboratory. Consult solubility tables to help predict whether an insoluble precipitate will accompany each reaction. Also predict whether a gaseous product might be produced. After completing your equations and your predictions, you will be assigned one of the reactions to be verified by observation. Be sure to use equal volumes of equal concentrations of hydrogen and hydroxide ion when you plan to neutralize an acid with a base. And remember that you can use litmus paper to verify that a volume of acid has been neutralized by a volume of base.

10-5. Precise Determination of Acid or Base Concentrations. The concentration of hydrogen ion and hydroxide ion in solutions can be determined very precisely by taking advantage of two properties of these ions. First, certain substances called *indicators* change color when very slight changes occur in the concentration of hydrogen ion. Second, the electrical conductivity of a solution may be affected sharply by changes in hydrogen ion concentration. In the case of indicators, litmus changes from blue to red when the concentration of hydrogen ion just exceeds the concentration of hydroxide ion. And phenolphthalein loses its pink color when the hydrogen ion concentration goes much above 1×10^{-9}M.

The determination of the concentration of an acid or base can be done by titration, a process whereby carefully measured volumes of two solutions are combined. Fundamental to the titration process is the fact that equal numbers of hydrogen and hydroxide ions combine to produce water molecules. Furthermore, the volume of a solution multiplied by the concentration of a dissolved substance, assuming the concentration is expressed in molarity, yields the number of moles of the dissolved substance in the solution. Combining these two bits of information produces the useful expression, at neutralization: Volume of acid $\times$ Molarity converted to normality $=$ Volume of base $\times$ Molarity converted to normality. The titration process enables us to measure volumes to tenths of a milliliter. Consequently, if the molarity of either the acid or base is known, the molarity of the unknown solution can be calculated.

To practice this kind of titration analysis, prepare 500 ml of 0.50 M sodium hydroxide solution. When preparing this solution, weigh the sodium hydroxide to centigrams and use a volumetric flask to establish the volume of the solution. Before transferring this solution to a burette, rinse a clean burette with several 15 ml portions of the solution. Check to be sure during each rinse that the tip is delivering the solution without bubbles or other sources of error.

For the acid solution, use a pipette to transfer a carefully measured volume of clear household vinegar to a 150 Erlenmeyer flask. Use less than a 10 ml sample. Add a drop or two of phenolphthalein solution to serve as an indicator, that is, to "indicate" when you have added as much OH ion as the vinegar contains H ion. Actually, at neutralization the concentration of hydrogen ion is 1×10^{-7}M but phenolphthalein changes color at approximately 1×10^{-9}M but no indicator is available which serves so conveniently to show concentration changes any closer to 1×10^{-7}M.

Consult the label on the vinegar container and calculate the concentration of hydrogen ion in this brand of vinegar. For example, a brand of vinegar that is 6% acetic acid contains approximately 60 g of acetic acid per liter. Since one mole of acetic acid weighs 60 grams, the vinegar is 1 M acetic acid. By way of this calculation, we know that we will need approximately twice the volume of our 0.5 M base solution to neutralize a specified volume of the vinegar. Consequently, the base solution can be added relatively rapidly until the neutralization volume becomes imminent. Approaching neutralization is indicated by how long the pink color persists as you whirl the flask during the titration process. Careful titration permits determining the neutralization point within a fraction of a drop of solution.

At neutralization, read the burette to determine the volume of the base used and perform the computations needed to determine the molarity of the acetic acid vinegar solution. To assure yourself of the precision of your titration, repeat the process using at least two more samples of the vinegar. Decide within what limits of precision you could "guarantee" the concentration of the vinegar.

INSTRUCTIONAL ACTIVITIES: TEACHER

Important though it is, the topic of acids and bases confronts the teacher with several concepts which seem to be very difficult for students to grasp. Furthermore, many of the existent instructional activities seem to focus attention on the more obvious leaving it up to the student to grasp the abstract, theoretical descriptions of acid-base phenomena primarily by the interpretation of text materials. Examples follow.

Students find difficulty in understanding the strategy of titration primarily in connection with the non-linear changes in hydrogen ion concentration in solutions approaching the hydrogen ion concentration in pure water. Apparently drastic adjustments must be made in students' minds when instructional vocabulary makes the transition between talking simply about bases versus "conjugate" bases and other aspects of the Bronsted-Lowry manner of describing acid-base interaction. Ideas which seem to make sense to students when illustrated by reference to mono-basic acids become increasingly confused when we shift to multi-basic acids and especially so when we apply K_A calculations to such acids.

In general, when approaching the topic of acids and bases seems to be an appropriate time to encourage teachers to sharpen up the precision and clarity of the instructional activities they choose to present the concepts they choose to teach and, if necessary, to defer to instructional levels beyond the high school those concepts for which adequate instructional techniques have not yet been developed.

10-T1. Indicators and the Strength of Acids and Bases. Prepare solutions of such indicators as orange IV, thymol blue, methyl orange, methyl red, bromcresol green, bromthymol blue, litmus, phenolphthalein, alizarin yellow R, and indigo carmine. Demonstrate the colors of each of these indicators in distilled water. Prepare 0.1 N solutions of such acids as hydrochloric, nitric, sulfuric, metaphosphoric, acetic, citric, tartaric, and boric. Invite the students to predict the hydrogen ion concentration in each of these acid solutions. After recording their predictions, place 10 ml samples in sets of small test tubes and add a drop or two of each of the indicators which are likely to change colors at concentrations of hydrogen ion in the range of 0.1 to 0.00005 M. Suggest that you will be willing to verify the 0.1 N concentrations of each of the acids by titrating 10 ml portions against 0.1 N base.

Having confronted your students with the observation that equal concentrations of different acids do not necessarily produce equal concentrations of hydrogen ion, interpret this to suggest that acids differ in degree of proton donor activity.

Prepare 0.1 N solutions of sodium hydroxide, potassium hydroxide, lithium hydroxide, ammonium hydroxide, and barium hydroxide. Demonstrate the colors of each of the indicators in each of these solutions. Draw tentative conclusions regarding the degrees of proton acceptance demonstrated by each of these bases.

Explore the idea that the color of an indicator might be used to determine whether the indicator particle in solution is in itself acting as a proton donor or acceptor. Cite such evidence as the high degree of probability that all of the indicators would be proton acceptors in 0.1 N acids and proton donors in 0.1 N bases. Examine the evidence provided by each indicator against the preceding hypothesis. Allow the analysis of the evidence to evolve toward suggesting that each indicator has its own "ability" to donate or accept protons and, thus shows acid-base behavior while serving as an indicator.

10-T2. The Bronsted-Lowry Concept of Acid-Base Interaction. Teachers differ widely in the degree to which they use metaphors, analogs, parables, and similar verbal efforts to help their students catch the general spirit or character of concepts. These differences are very probably quite closely correlated with highly personal attributes and characteristics. And one teacher's favorite metaphor or analog may, to another equally motivated and scholarly teacher, seem to be total nonsense or even downright misleading. But we find ourselves rather often saying something like, "What it is that is puzzling you is somewhat like. . ." On this basis, teachers should now and then look at the metaphors, analogs, and other similar verbal constructs they use to examine their validity and appropriateness.

Consider, for example, the following effort to help students understand and appreciate the Bronsted-Lowry concept of acid-base interaction. Were a very rich man to meet a very poor man there is little question as to which of the two men should give and which should receive a hand-out. But what would be the state of affairs were this rich man to meet a much, much richer man or this poor man a much, much poorer man? And here is a more involved question. Suppose a very poor man were to meet a very rich man at the same moment the very rich man was meeting a man of middle-class wealth. Would the very poor man make out as well upon meeting the two men simultaneously as he might were he to meet the two richer men separately?

Having entertained a discussion of these metaphorical questions for a few minutes, confront your students with 0.1 N solutions of the acids used in 10-T1. Demonstrate the proton donating tendency of each acid by using pH paper or a pH meter. Invite predictions of the pH's of solutions obtained by combining equal volumes of various combinations of the acids. Be sure to include combinations involving wide ranges in K_A values. After predictions have been recorded, perform the indicated combinations and determine the pH's of the resulting solutions. Invite your students to assay the validity of the "rich man-poor man" metaphor in light of the demonstrated pH's of these solutions.

Substitute the series of bases used in 10-T1 and go through the same general strategy.

Finally, confront your students with the proposition that you are about to combine equal volumes of selected acids with selected bases and invite their predictions of the pH's of the resulting solutions. Having recorded their predictions, proceed with performing the indicated combinations and determine the pH's of the resulting systems. These results will very probably leave many students in a state of puzzlement. Assuming that this state of affairs coincides with the end of the day's lesson, invite your students (for homework) to be ready to attack or defend the following hypothesis: If the solutions resulting from combining equal volumes of equal normality acids and bases are evaporated and their dissolved salts are put back into solution, all solutions so prepared would have equal pH values.

On the following day, devote adequate time to allowing your students to argue the merits of their individual attacks or defenses of this hypothesis. Arrange to have yesterday's solutions evaporated during the night and be ready to perform the operations needed to verify or disprove or, more probable, provide evidence that strengthens or weakens the various stands taken by your students.

10-T3. Acid-Base Interactions and the Buffering Phenomenon. Using the K_A for acetic acid, calculate the pH (chalkboard) of 0.01 N acetic acid. Similarly and assuming no interaction between sodium acetate and water, calculate the pH of a solution of 0.01 M sodium acetate. Again assuming no interactions other than possible neutralization, calculate the probable pH of a solution prepared by combining equal volumes of 0.01 M acetic acid and sodium acetate. Contrast the value derived by these calculations made under the stated assumptions with the demonstrated (4.75) pH of the indicated solution.

Announce that before looking into this apparently puzzling state of affairs, you would like to provide an additional observation. For the preliminary step, calculate the pH of a solution that might be prepared by combining equal volumes of the acetic acid sodium acetate and 0.001 M hydrochloric acid. Again, contrast the value derived by this calculation made under the stated assumption with the demonstrated pH of the indicated resulting solution (4.67).

Invite tentative explanations of the apparent discrepancies between the calculated values, reemphasizing the assumptions under which they were made, and the observed values. If responses seem to be too sluggish or ranging too far afield, be ready to suggest the equilibrium between undissociated acetic acid and its ion products and the relationships described by the Le Chatelier Principle.

10-T4. A Computerized Lesson Involving Titration Curves. The availability of computer facilities can invite chemistry teachers to look into their use when a lesson involves burdensome computations. A good example is the plotting of acid-base titration curves. The programming of this kind of lesson is illustrated in the following instructional activity.

In the first example, titled CHEM 1, the computer was "told" that the titration would begin with 2.5 ml of 0.1 N acid and 0.1 N base would be added drop by drop (0.05 ml). The computer was also "told" how to calculate the pH of the initial 2.5 ml of 0.1 N acid after each drop of the base was added and to carry out these calculations successively until 10 drops of the base had been added. And the accompanying read-out was to show the changing pH and the corresponding decrease in hydrogen ion concentration as moles of hydrogen ion.

The program follows:

```
OLD PROBLEM NAME—CHEM1*
WAIT.
READY.
LIST
CHEM1*   10:14   WSH   TUE   2/21/67
90 PRINT "TYPE VOLUME, ACID CONCENTRATION, BASE
  CONCENTRATION"
100 INPUT V,C1,C2
200 LET M=V*C1*.001
210 LET D=.05*C2*.001
215 PRINT
220 PRINT "NO. DROFS","PH VALUE","MOLES"
230 GOSUB 500
```

```
300 FOR N = 1 TO 1000
310 LET M=M—D
320 IF ABS(M*100)<D THEN 520
330 IF M<0 THEN 540
335 GOSUB 500
340 IF ABS(M)>=V*C1*.001 THEN 999
350 NEXT N
500 PRINT N,—LOG (M/(.001*V))/LOG(10),M
510 RETURN
520 PRINT N,"7.0","0"
530 GOTO 350
540 PRINT N,14+LOG(ABS(M)/(.001*V))/LOG(10),M
550 GOTO 340
999 END
```

Rather than showing the read-out for this program with the values computed for all 100 steps, drop by drop, the data show the computations for the first 5 drops, then drops 44 through 55, and, finally, drops 95 through 100.

```
OLD PROBLEM NAME—CHEM1*
READY.
RUN
CHEM1* 10:26 WSH TUE 2/21/67
TYPE VOLUME,ACID CONCENTRATION,BASE CONCEN-
  TRATION
? 2.5, .1,.1
```

NO. DROPS	PH VALUE	MOLES
0	1.	.00025
1	1.00877	.000245
2	1.01773	.00024
3	1.02687	.000235
4	1.03621	.00023
5	1.04576	.000225
44	1.92082	.00003
45	2.	.000025
46	2.09691	.00002
47	2.22185	.000015
48	2.39794	.00001
49	2.69897	.000005
50	7.0	0

51	11.301	—.000005
52	11.6021	—.00001
53	11.7782	—.000015
54	11.9031	—.00002
55	12.	—.000025
95	12.9542	—.000225
96	12.9638	—.00023
97	12.9731	—.000235
98	12.9823	—.00024
99	12.9912	—.000245
100	13	—.00025

This program can be modified to bring out even more dramatically how the pH of a solution changes during titration as the end-point is reached. Although the modified program calls for five times as many steps, the new program "tells" the computer to increase the volume of base being added beyond the single drop quantity if the pH is not changing significantly. Similarly, if the pH is changing more rapidly than a predetermined rate, the computer is "told" to diminish the volume of each aliquot. The program follows.

```
OLD
OLD PROBLEM NAME—CHEM2*
READY.
LIST
CHEM2*  10:22  WSH  TUE  2/21/67
100 REM DEFINE PH FUNCTION
110 DEF FNA(X)=LOG(ABS(X)/(.001))/LOG(10)
120 PRINT "PH TITRATION CURVE FOR UNIPROTIC
    ACID WITH ADDED BASE"
130 PRINT
140 REM ASK FOR INPUT INFORMATION
150 PRINT "WHAT IS THE AMOUNT AND CONCENTRA-
    TION OF THE ACID";
160 INPUT V1,C1.
170 PRINT "WHAT IS THE CONCENTRATION OF THE
    BASE";
180 INPUT C2
190 REM V2 IS THE AMOUNT OF BASE ADDED EACH
    DROP
200 LET V2=.05
```

```
210 REM M1=MOLES IN SOLUTION
220 LET M2=C1*V1*.001
230 REM M2=MOLES PER DROP
240 LET M2=C2*V2*.001
250 LET N=0
260 PRINT
270 PRINT "PH", "NO. OF DROPS", "MOLES(+=ACID,-
    =BASE)"
280 PRINT
290 GOSUB 760
300 REM PRINT RESULTS OF THE 1ST 10 DROPS
310 LET J=1
320 LET X1=10
330 LETX2=1.3
340 GOSUB 820
350 REM CHANGE SCALE FOR PRINTING RESULTS
360 LET L=.2
370 GOSUB 960
380 LET X1=1000
390 GOSUB 820
400 REM CHANGE SCALE
410 LET X2=2
420 GOSUB 960
430 GOSUB 820
440 REM CHANGE SCALE
450 LET X2=3
460 GOSUB 960
470 GOSUB 820
480 REM CHANGE SCALE
490 LET J=1
510 LET X2=11
520 GOSUB 820
530 PRINT "7.0",N,"0"
540 REM CHANGE SCALE
550 GOSUB 890
560 LET L=L*4
570 GOSUB 960
580 LET X2=11.7
590 GOSUB 960
580 LET X2=11.7
590 GOSUB 890
600 REM CHANGE SCALE
```

```
610 GOSUB 960
620 LET X2=12.2
630 GOSUB 890
640 REM CHANGE SCALE
650 GOSUB 960
620 LET X2=12.2
630 GOSUB 890
640 REM CHANGE SCALE
650 GOSUB 960
660 LET X2=14-LOG(1/C1)/LOG(10)
670 GOSUB 890
680 STOP
690 REM ADD DROPS
700 FOR I=1 TO J
710 LET M1=M1-M2
720 LET N=N+1
730 NEXT I
740 RETURN
750 REM PRINT ACID PH
760 PRINT -FNA(M1),N,M1
770 RETURN
780 REM PRINT BASE PH
790 PRINT 14+FNA(M1),N,M1
800 RETURN
810 REM TEST FOR SCALE CHANGE
820 FOR K=1 TO X1
830 IF -FNA(M1)>=X2 THEN 880
840 GOSUB 700
850 IF ABS(M1*100)<M2 THEN 530
860 GOSUB 760
870 NEXT K
880 RETURN
890 REM TEST FOR SCALE CHANGE
895 FOR K=1 TO X1
900 IF 14—FNA(M1)>=X2 THEN 940
910 GOSUB 700
920 GOSUB 790
930 NEXT K
940 RETURN
950 REM GENERATE NEW SCALE VALUE
960 LET J=INT(ABS(M1)/M2*L)
970 IF J=0 THEN 990
```

```
980 RETURN
990 LET J=1
1000 GOTO 980
1010 END
```

This and the preceding program were developed with several very specific ideas in mind. Assuming the student was already familiar with the calculations which accompany the introductory titration exercises, the pattern in these calculations should show through the computer language. Thus, the student can follow the transition from hand to computerized calculations. Furthermore, the student knows how long it takes to do these calculations by hand, a length of time in dramatic contrast to the five seconds needed by the computer to produce the following readout.

```
OLD
OLD PROBLEM NAME—CHEM2
READY
RUN
CHEM2*  10:18  WSH  TUE  2/21/67
PH TITRATION CURVE FOR UNIPROTIC ACID WITH
  ADDED BASE
WHAT IS THE AMOUNT AND CONCENTRATION OF THE
  ACID? 2.5,.1
WHAT IS THE CONCENTRATION OF THE BASE? .1
PH        NO. OF DROPS   MOLES[+=ACID,-=BASE]
 1.        0              .00025
 1.00877   1              .000245
 1.01773   2              .00024
 1.02687   3              .000235
 1.03621   4              .00023
 1.04576   5              .000225
 1.05552   6              .00022
 1.0655    7              .000215
 1.07572   8              .00021
 1.08619   9              .000205
 1.09691   10             .0002
 1.18046   17             .000165
 1.284     24             .00013
 1.42022   31             .000095
 1.49485   34             .00008
```

1.58503	37	.000065
1.69897	40	.00005
1.85387	43	.000035
2.09691	46	.00002
2.22185	47	.000015
2.39794	48	.00001
2.69897	49	.000005
7.0	50	0
11.301	51	—.000005
11.6021	52	—.00001
11.7782	53	—.000015
12.	55	—.000025
12.1461	57	—.000035
12.2553	59	—.000045
12.5052	66	—.00008
12.6628	73	—.000115
12.7782	80	—.00015
12.8692	87	—.000185
12.9445	94	—.00022
13.0086	101	—.000255

TIME: 5 SECS.

10-T5. Computer-Assisted Instruction Involving Simulated Titration. Efforts are being explored to combine instruction and computational facilities in "self-contained" computer-assisted chemistry lessons. For example, John M. Boblick has developed an experimental lesson in which the student not only uses the computer to do the required computations but also to provide the instruction needed to allow the student to practice the general techniques for determining the strength of an acid by titration with a standardized base. In his report to the Montgomery County, Maryland, Computer-Assisted Instruction Demonstration Project, Boblick describes the lesson as a "fast, efficient method for students to practice the application of the chemical principles associated with titration."

Before the student sits down at his computer-assisted instruction station, he receives a pre-lab instruction sheet with such information as:

> This lesson is designed to give you practice in performing an acid-base titration. The program will permit you to add, in small

increments, a standardized solution of NaOH to an acid of unknown hydrogen ion concentration.

After the acid has been neutralized, the program will display for you the:

(1) ml of acid used;
(2) ml of base used;
(3) molarity of the base.

With this data you will be able to calculate the hydrogen ion concentration of the acid.

The apparatus depicted within this simulation consists of a 50 ml burette and a 250 ml flask. The program will randomly select an acid of unknown strength (between 0.1 and 5.0 molar). You may choose the volume (ml) of acid which is added to the flask.

After selecting the molarity of the standard NaOH solution which you wish to use, you will titrate the NaOH solution from the burette. Following the addition of each portion of base, you will be informed if the resulting solution is acidic, basic, or neutral. If the solution is still acidic, more base must be added. If the solution becomes basic, it will be necessary for you to back-titrate; i.e., add small amounts of acid to the flask. Continue to add smaller and smaller amounts of acid or base (whichever is required) to the flask until the solution is neutralized.

By combining instructional "dialog" with computer capabilities, the program provides immediate feedback of the accuracy of the student's work. The opening and closing "frames" of the actual program illustrate the format. The program is written in APL (a Programming Language).

AN ACID-BASE TITRATION

John M. Boblick, author. February, 1969

You are to determine the hydrogen ion concentration of an unknown acid by titration with a standard base (NaOH). You will be able to determine the volume of base required to neutralize the volume of acid which you use. To continue, press the space bar.

After determining the volume of base needed to neutralize the acid used, you will be placed in the computational mode so that you may use the data collected to compute the hydrogen ion concentration of the acid. To continue, press the space bar.

What volume (ml) acid do you wish to use? □ (Student response). 10

Enter the molarity of the base (NaOH). □ 1.0.

What volume (ml) base do you wish to use? □ 5.

For 5 ml base added, the solution is acidic. Add more base. What volume (ml) base do you wish to use? □ 5 .....

.....For 35.5 ml base added, the solution is acidic. Add more base. What volume (ml) base do you wish to use? □ 0.1.

Solution is basic. You have passed the end point. You must back-titrate by adding more acid. What volume (ml) acid do you wish to use? □ 0.0

Your titration record is as follows: ml acid used = 11.1
ml base used = 35.6
molarity of base = 1.0

To calculate the hydrogen ion concentration of the acid, signal *attention;* to continue, enter →92.

G(91) (Student calculation) 35.6 × 1.0 ÷ 11.1 = 3.20721

→92 The hydrogen ion concentration is 3.2 moles/liter.

When you start again, you will have a new acid to work with.

As Boblick points out, acid-base titration using burettes is no longer as important as a laboratory tool as it was before the days of recording pH meters and other modern instruments. The above program, however, provides a fast, efficient medium for the study of acid-base reactions and increases student interest by taking advantage of the motivation that seems to be generated by computer-based simulations.

Audio-Visual Aids

10-AV1. "Acid-Base Indicators." Color. 19 min. Chem Study. Indicator effect as proton donor-acceptor action. Hydrogen ion concentration effect on and equilibrium constants of indicators.

10-AV2. "Acid-Base Theories." 40 frames. FOM. Arrhenius and Bronsted concepts developed by experiments and models. Conjugate acid-base table explained. Indicators and pH.

10-AV3. "Acids." Color. 5 min. Disraeli Films. Safety precautions in pouring acids.

10-AV4. "Acids, Bases, and Oxides." Color. 14 min. Purdue University. Preparation and properties.

10-AV5. "Acids, Bases, and Salts." Color. 21 min. Coronet. Properties, theories, preparation, and uses.

10-AV6. "Chart of Relative Strengths of Acids and Bases." 42 × 58 inches. Welch Scientific. Equilibrium constants.

10-AV7. "Handling Liquids." Color. 4 min. Yale University. Good versus bad practices. How to identify acids before opening containers.

10-AV8. "Indicators." Color. 28 min. Coronet. Using methyl violet and methyl orange to determine ionization constant of acetic acid.

10-AV9. "Ion Removal by Metathesis." Color. 4 min. Yale University. Phenomena associated with acid-base interaction.

10-AV10. "Properties of Acids, Bases, and Salts." Color. 28 min. Yale University. Acids and bases as solutions of metallic and non-metallic oxides.

10-AV11. "Standard Solutions and Titration." Color. 21 min. Coronet. Titration techniques and calculations using sodium hydroxide and potassium hydrogen phthalate.

10-AV12. "Sulfuric Acid." Color. 16 min. Coronet. Physical and chemical properties. Handling. Effects of concentrated versus dilute.

Ideas for Additional Instructional Activities

10-AD1. Atkinson, R. C., and H. A. Wilson, "Computer-Assisted Instruction," *Science,* 162: 73–77. October 4, 1968.

10-AD2. Breedlove, C. H., Jr., "Determination of Acetic Acid in Vinegar Using Gravimetric Titrimetry," *Chemistry,* 42: 24–25. May, 1967. Using the balance to improve precision.

10-AD3. Darnowski, Vincent S., "Computer-Aided Instruction, a Tool for Science Teaching in the Seventies," *The Science Teacher,* 35: 22–28. January, 1968.

10-AD4. Ellis, R. H., and E. R. Linner, *Journal of Chemical Education,* 26: 528–529. 1949. Gravimetric investigation of reactions between metals and acids.

10-AD5. Gillespie, R. J., "Cryoscopy—What We Can Learn from Freezing Points," *Chemistry,* 38: 12–18. October, 1965. Using sulfuric acid as the solvent.

10-AD6. Hammer, A. J., "Simultaneous Standardization of 0.1 Hydrochloric Acid and Sodium Hydroxide Using Calcite," *Journal of Chemical Education,* 11: 244. 1934. By adding a known mass of calcite to an excess of the acid and then titrating with the base. Representative data are included.

10-AD7. Kokes, R. J., M. K. Dorfman, and T. Mathia, "Experiments for General Chemistry: III, Calorimetry," *Journal of Chemical Education,* 39: 90. 1962. Enthalpies of neutralization.

10-AD8. Lapp, Walter S., and Edgar T. Wherry, "pH Preferences of

Common Plants," *The Science Teacher,* 18: 121–126. April, 1951. Testing soil pH and optimum levels for selected plants.

10-AD9. Morris, L. H., "The Reaction Between Water and Magnesium," *School Science Review,* 43: 451. 1962. Compares rates of reaction between magnesium powder and distilled versus tap water.

10-AD10. Ronneberg, C. E., "A Simple Procedure for the Determination of the Equivalent Weight of Magnesium," *Journal of Chemical Education,* 14: 71. 1937. Includes discussion of possible errors in procedure whereby the volume of hydrogen liberated is measured directly.

10-AD11. Shan, N. M., *Journal of Chemical Education,* 12: 492–493. 1935. Compares volume of water displaced by known volume of hydrogen versus volume of hydrogen collected over water.

10-AD12. Shen, Samuel Y., and Alfred J. Gilman, "Potentiometric Titration of Aspirin in Ethanol," *Journal of Chemical Education,* 42: 540. October, 1965. Based on production of acetyl salicyclic acid.

10-AD13. State, Harold M., "Writing a Chemical Equation from Titration Data," *Journal of Chemical Education,* 39: 297. 1962. Knowing grams of solute per ml of solution, students determine the equations for such reactions as phosphoric acid and sodium hydroxide or iodine and sodium thiosulfate.

10-AD14. Stock, John T., "The pH Indicator," *The Science Teacher,* 29: 28–31. November, 1962. pH and its measurement.

10-AD15. Strong, Laurence E., L. B. Clapp, and J. O. Edwards, "Geometry in the Beginning Chemistry Course," *Journal of Chemical Education,* 38: 530. 1961. Includes discussion of formation of hydronium and chloride ions when hydrogen chloride is dissolved in water.

10-AD16. Stong, C. L., "Making an Apparatus That Will Measure the Acidity or the Alkalinity of Solutions," *Scientific American,* 219: 232–247. September, 1968. Includes general discussion of pH.

10-AD17. Szabadvary, Ferenc, "Indicators: a Historical Perspective," *Journal of Chemical Education,* 41: 285–287. May, 1964. Natural as well as synthetic. Theory of indicator action.

10-AD18. Ts'ao, Ching-Pan, "Determination of Equivalent Weight of Metals," *Journal of Chemical Education,* 17: 598. 1940. Reacting magnesium, aluminum, or zinc with acid and collecting hydrogen by displacement of water.

10-AD19. Turner, A. Mason, "A New Approach to Acids, Bases, and Salts," *The Science Teacher,* 35: 62–65. December, 1968. Step by step instructions.

10-AD20. Zuehlke, R. W., "Laboratory Group Exercises in Acid-Base

Theory," *Journal of Chemical Education,* 39: 354. 1962. Combines indicator theory with titration skills.

10-AD21. Zuder, W. F., W. S. McGuire, and S. Zuffanti, "Teaching the Electronic Theory of Acids and Bases in the General Chemistry Course," *Journal of Chemical Education,* 20: 344. 1943. Lecture demonstrations. Includes acid-base reactions in water, anhydrous chlorobenzene, and anhydrous carbon tetrachloride.

11

OXIDATION—REDUCTION

To beginning students, oxidation may be "another name for burning" or a "process in which oxygen combines with another substance." And the term, "reduction," may have no chemical connotations. But as the chemistry course unfolds, we expect our students to associate these terms with the abstract concepts of electron transfer and the accompanying energy transformations.

It is natural to emphasize the importance of oxidation-reduction reactions. Electron transfers are at the very center of many reactions. The mutual attractions of atomic nuclei for one or more electrons is pretty much the "glue" that holds the universe together. Ideally, students will come to realize that to keep track of electron transfers provides a vantage point from which to describe a chemical event. And whether a reaction will or won't occur can be predicted more accurately if it is known which particles are more likely to hang onto or lose electrons when they bump into each other.

Some teachers choose to combine the topic of this chapter with electrochemistry. The logic is that students are interested in "wet" and "dry" cells and the many uses they are put to in today's technology. Beginning with the translocation of electrons through external circuits, instruction can proceed to focus attention on the internal reshufflings of particles which give rise to electron flow. Supposedly, this approach improves the perspective of students as they become involved with "oxidation numbers," "half-reaction equations," and other ideas used to describe oxidation-reduction reactions.

INSTRUCTIONAL ACTIVITIES: STUDENT

11-1. Competition for Electrons. Collect small pieces of polished or freshly cut zinc, magnesium, iron, copper, lead, tin, calcium, and additional metals if available. In as many different test tubes as you have different kinds of metals, put approximately 10 ml of 1.0 M hydrochloric acid. Prepare a second set of test tubes containing 1.0 M nitric acid. Carefully (*hood*) drop a metal sample in each of both sets of test tubes. If no reaction is evident in some of the test tubes, warm gently but do not boil.

Looking forward to interpreting your observations, recall the following bits of information: (1) hydrochloric acid contains hydrogen and chloride ions and nitric acid contains hydrogen and nitrate ions, (2) nitrate ions in the presence of hydrogen ions and excess electrons can produce nitrogen dioxide and water, (3) atoms of a metal in the solid, undissolved state contain equal numbers of electrons and protons but when metals go into solution they are likely to lose one or more electrons per atom, (4) it is difficult to think of electrons as existing free and unattached to some kind of particle although electrons can pass "baton-wise" from particle to particle, (5) in water environments, it is also difficult to think of hydrogen ions (protons) as existing free and unattached to other particles, especially the negative portions of adjacent water molecules.

Examine each test tube and decide whether or not the metal has gone into solution. Arrange each set of test tubes in the order of increasingly rapid reactions and record this order. The next step in this activity explores the relative abilities of the metals to hang onto their electrons. For example, you very probably found that the hydrogen ions in the hydrochloric acid took electrons away from zinc atoms and the zinc went into solution and the nitric acid system caused copper atoms to lose electrons and go into solution. What happens if a piece of copper is put into a solution of zinc ions? Will the zinc ions regain the electrons they lost by taking electrons from the copper and, thus, cause the copper to go into solution?

Conversely, if a piece of zinc is dropped into the solution of copper ions, does the copper reappear as "coppery" metal and the zinc go into solution?

Plan an investigation that will enable you to list all of your metals in the order of their decreasing ability to hang onto their electrons. You will need to divide your solutions so as to test all of the metals in each of the solutions or you may choose to team up with other class members and pool your results. Ions of which metal seem to be able to regain electrons from all of the other metals? In other words, if oxidation is loss of electrons, which metallic ion can oxidize all of the other metals? If reduction is the gain of electrons, which metal can reduce the ions of all of the other metals?

Assuming that the earth's oceans and lakes contain at least small concentrations of all of the common metal ions, which metals are least likely to be found in nature in the pure, elemental form?

11-2. Competition for Electrons by the Members of the Halogen Family. Recall Instructional Activity 8-T3 and, if necessary, repeat the procedure of the demonstration as a laboratory exercise. Interpret all of your observations from the perspective of competition for electrons. In other words, decide which member of the halogen family is least able to hang onto extra electrons and, thus, exist as ions. Similarly, which halogen molecule will "steal" electrons from the ions of all of the other members of the family?

INSTRUCTIONAL ACTIVITIES: TEACHER

Sometimes a relatively large array of observations and related information must be derived and processed "in one sitting" else students lose the evolving thread of interrelationships. Furthermore, class sessions of fifty minutes or so or even double period laboratory sessions always seem to come to an end before laboratory derived observations can be fully analyzed and the accompanying principles or concepts fully synthesized. The following instructional activity illustrates the use of a "guide sheet" to speed along a fully packed demonstration.

11-T1. Step-by-Step Oxidation and Reduction. Announce that the day's presentation involves an extensive array of observations to be derived and processed if the "punch line" is to be achieved and that you have, therefore, prepared a guide sheet to help them organize their work. (Note: Many students have acquired a negative attitude toward any kind of "fill-in-the-blanks" instructional format and, hence, need to be sold on the efficiency aspect of the guide sheet.) You may or may not choose to reveal the subtle motive of helping the students appreciate the use of equations as a convenient and effective means of communication.

As background information useful to interpret your observations during the presentation, in the spaces provided on the guide sheet record your idea of the color and other characteristic properties of the following systems:

> iron(II), iron(III), iodide ion and trichlorotrifluoroethane (TTE), iodine water and TTE, manganese(II), permanganate ion, manganate ion, mercury(II), chromium(III), dichromate ion, chromate ion, iron(II) and ferricyanide ion, and iron(III) and ferricyanide ion.

Your guide sheet indicates a series of systems which we will now prepare. I will identify each system but, since you can observe the reaction or lack of reaction, you are to hypothesize the products of each reaction. Don't forget that the lack of an apparent reaction also provides useful information when we are trying to establish the relative tendencies for particles to take on or give up electrons.

1. Iron(II) + iodide ion + TTE $\longrightarrow$
2. Iron(II) + iodine water + TTE $\longrightarrow$
3. Iron(III) + iodide ion + TTE $\longrightarrow$
4. Iron(III) + iodine water + TTE $\longrightarrow$

Pause at this point and invite the students to identify the action being featured in this set of four systems. Nudge the students toward suggesting that you add ferricyanide ion to the four test tubes to shed additional light on one or another hypothesis. If you are using this instructional activity more to "clarify" than to "discover" relative E° values, you may include in their guide sheet the following portion of an E° table:

$Mn \longrightarrow 2e + Mn^{2+}$	1.18 volts
$Cr \longrightarrow 3e + Cr^{3+}$	.74
$Fe \longrightarrow 2e + Fe^{2+}$	.44
$Cr^{2+} \longrightarrow e + Cr^{3+}$	.41

$H_{2(g)} \longrightarrow 2e + 2H^+$	0.00
$2I^- \longrightarrow 2e + I_2$	− .53
$MnO_4^{2-} \longrightarrow e + MnO_4^-$	− .56
$Fe^{2+} \longrightarrow e + Fe^{3+}$	− .77
$NO_{2(g)} + H_2O \longrightarrow e + NO_3^- + 2H^+$	− .78
$Hg \longrightarrow 2e + Hg^{2+}$	− .78
$2Hg_2^{2+} \longrightarrow 2e + 2Hg^{2+}$	− .90
$NO_{(g)} + 2H_2O \longrightarrow 3e + NO_3^- + 4H^+$	− .96
$Mn^{2+} + 2H_2O \longrightarrow 2e + MnO_2 + 4H^+$	−1.28
$2Cr^{3+} + 7H_2O \longrightarrow 6e + Cr_2O_7^{2-} + 14H^+$	−1.33
$Mn^{2+} + 4H_2O \longrightarrow 5e + MnO_4^- + 8H^+$	−1.52

For the second set of systems to be observed by the students, display test tubes containing solutions of MnO_4^-, Mn^{+2}, MnO_4^{2-}, and a suspension of MnO_2. Follow this by confronting the students with systems containing approximately equal volumes of 0.1 M solutions:

5. $MnO_4^- + H_2O_2 \xrightarrow{H^+}$
6. $MnO_4^- + H_2O_2 \xrightarrow{OH^-}$
7. $MnO_4^- + Fe^{2+} \longrightarrow$
8. $MnO_4^- + Fe^{2+} \xrightarrow{H^+}$
9. $MnO_4^- + Fe^{2+} \xrightarrow{OH^-}$
10. $MnO_4^- + KNO_{3(aq)} + KOH_{(aq)} \xrightarrow{\text{Heat}}$
11. $MnO_4^- + KI \longrightarrow$
12. $MnO_4^- + KI \xrightarrow{H^+}$
13. $MnO_4^- + KI \xrightarrow{OH^-}$

All observations should be studied against the reference solution and discussed from the point of view that certain particles may have gained or lost electrons between the initial and final states of each system. Prior to the discussion, however, each student should be given an uninterrupted opportunity to record his idea of the products from each reaction and to balance his hypothesized equations for these reactions.

This instructional activity can be extended by substituting $Cr_2O_7^{2-}$ ions for the MnO_4^-. The reference set of test tubes should display ionic systems of $Cr_2O_7^{2-}$, CrO_4^{2-}, and Cr^{3+}.

Audio-Visual Aids

11-AV1. "Chromium and Manganese." Color. 37 min. Coronet. Features oxidation-reduction reactions.

11-AV2. "Metals and Non-Metals." Color. 11 min. Coronet. Physical and chemical properties, especially roles in oxidation-reduction reactions. Alloys.

11-AV3. "Oxidation." 44 frames FOM. Experiments of Lavoisier. Combustion.

11-AV4. "Oxidation and Reduction." 11 min. Britannica. At the general level of burning of phosphorus and rusting of iron.

11-AV5. "Oxidation-Reduction." 40 frames. FOM. Rules for determination of oxidation state and oxidation number. Various methods for balancing redox reactions include electron shift, ion-electron, and half cell reactions.

11-AV6. "Oxidation-Reduction." Color. 9 min. McGraw-Hill. In terms of electron transfer.

11-AV7. "Oxidation-Reduction Half-Reactions Chart." 42 $\times$ 58 inches. Welch Scientific. Oxidation potentials for the half reactions most commonly encountered in high school chemistry.

11-AV8. "Preparation and Properties of Nitric Acid." Color. 21 min. Coronet. Preparation from sodium nitrate and sulfuric acid. Oxidation and reduction properties.

11-AV9. "The Story of Alfred Nobel." 11 min. Films. His work and his decision to establish the Nobel Award Program.

Ideas for Additional Instructional Activities

11-AD1. Alyea, Hubert N., and F. B. Dutton, *Tested Demonstrations in Chemistry*. Easton, Pennsylvania: *Journal of Chemical Education,* 1960. Section 8, pages 19–20, suggests several demonstrations involving oxidation and reduction phenomena. Page 56 describes a cell in which the reaction between H_2SO_4 and $KMnO_4$ can be investigated electrochemically. An investigation of oxidation and reduction of ions at electrodes is described on page 57. A colorful demonstration involving the Fe(III), $Fe_3[Fe(CN)_6]$, and FeCNS ions is described on page 63.

11-AD2. Fabishak, Vernon L., "Matrix Method for Balancing Chemical Equations," *Chemistry,* 40: 18–21. December, 1967. Correlates chemical and mathematical concepts.

11-AD3. Ferguson, Harold W., "Algebra for Balancing Redox Reactions," *Chemistry,* 40: 18–20. February, 1967. Avoids use of arbitrarily assigned oxidation numbers.

11-AD4. Gorman, M., "Some Electrochemical Experiments for Freshmen," *Journal of Chemical Education,* 34: 409. 1957. For example, the use of couples to study oxidizing and reducing agents.

11-AD5. Haas, Rudy, and Karl H. Gayer, "Balancing Ionic Equations by the Method of Undetermined Coefficients," *Journal of Chemical Education,* 39: 537. October, 1962.

11-AD6. Lauren, Paul M., "Lead Trees, Chemical Reactions, and Electron Transfer," *Chemistry,* 37: 23–26. December, 1964. Factors which influence crystal growth in silica gel.

11-AD7. Martin, D. R., "Lecture Demonstrations of Electrochemical Reactions," *Journal of Chemical Education,* 25: 495. 1948. Voltaic and electrolytic cells.

11-AD8. Meloan, Clifton E., "Balancing Oxidation Reduction Equations," *Chemistry,* 37: 12–18. July, 1964. Examples and cautions to consider when using such methods as (1) oxidation numbers, (2) half cell, (3) polar number, (4) hypothetical free radical, and (5) algebraic.

11-AD9. Newton, T. W., "The Kinetics of Oxidation Reduction Reactions." *Journal of Chemical Education,* 45: 571–575. Distinguishing various types of oxidation reduction reactions.

11-AD10. Pattison, D. B., J. B. Miller, and W. W. Lucasse, "Simplified Calorimetric Studies of Various Types," *Journal of Chemical Education,* 20: 319. 1943. The reaction between potassium bromate(V) with hydrobromic acid is studied from the double perspective of oxidation-reduction and enthalpy.

12

ELECTROCHEMISTRY

Today's students are familiar with the things dry cells and batteries can do. They may be less familiar with the chemical reactions which occur inside these electrochemical devices. Even so, the everyday applications of electrochemistry may serve for motivation when we introduce the topic.

Mental models of atomic structure are fundamentally abstract. The effects of certain electrical phenomena can be observed. On this basis, sometimes electrical phenomena provide a relatively concrete approach to the highly abstract domain of atomic structure. The flow of electrons through an external circuit can be traced to the reshuffling of atoms from molecule to molecule within the chemical system from which the electron flow occurs.

Electrochemical phenomena also provide opportunities for students to see or measure things which point to the interdependence among the various forms of energy. It is axiomatic that the world's demands for available energy increase rapidly, an increase that most certainly suggests that students of today will be called upon to make important decisions involving the conservation or development of energy sources as citizens of tomorrow.

INSTRUCTIONAL ACTIVITIES: STUDENT

12-1. Quantitative Description of Electron Holding or Electron Taking Properties of the Elements. If you recall Instructional Activity 11-1, you know that atoms of the different elements differ in the abilities to hold on to their own or to attract electrons from atoms or other elements. The goal of this activity is to measure this ability. Near the end of the activity we will identify a second goal that deals primarily with your ability to handle the vocabulary of chemistry.

The energy that causes electrons to move can be measured in units called volts—units measured by voltmeters. But it is not easy to arrange a system in which we can measure directly the electron holding or attracting property of a single element. It is more practical to arrange a system that compares the degree to which two elements exhibit this property.

For example, we can place a sheet or strip of copper in a solution of an acid that we know will cause copper atoms to lose electrons and become ions of copper. Similarly, we can put a strip of silver in the same kind of acid and some of the silver atoms will lose electrons and become silver ions. But suppose we arrange a system in which one metal can lose electrons only if ions of another metal will take them on. This arrangement will reveal valuable information about the relative electron attracting properties of the two metals. Which way the electrons flow identifies the "loser" versus the "taker." If we put a voltmeter in the circuit, we can measure the degree to which the electron attracting property of one metal exceeds that of the other.

Equipment can easily be developed that will solve the problem of comparing the electron attracting forces of two metals. There are some assumptions built into this set-up. For example, we assume that electrons cannot be lost by one and taken on by the other metal's ions without traveling through the external circuit. Also, a voltmeter across this external circuit will indicate the urge to flow on the part of the mobile electrons. There is also the assumption that a measurement of the voltage in the external circuit provides a valid description of the electron transfer or movement in the internal portion of the circuit.

In the case of a silver-copper cell, if the voltmeter reads 0.46 volt, we can say that the two metals, copper and silver differ by 0.46 volt in the electron attracting property. Furthermore, the voltmeter shows that the direction of electron flow is toward the silver in the external portion of the circuit and, consequently, away from the copper. Looking at the total system from the point of view of individual atoms, apparently copper atoms are losing electrons which travel through the external circuit to the silver metal and, thence, to the silver ions surrounding the silver metal. So long as the flow of electrons is maintained, copper atoms will go into solution as copper ions and silver ions will take up the electrons lost by the copper and plate out as new silver atoms on the silver metal.

Follow the instructions given by your teacher and compare the electron attracting property of several other pairs of metals. Usually, the instructions for these comparisons call for 1.0 M solutions of the ions of each of the two metals, freshly polished strips of the metals, and the separation of the two "half-cells" by an ion bridge or porous cup. Compare the readings on your voltmeter with the values published by experienced chemists using highly controlled procedures and precision instruments.

While reading this exercise, you may have felt that the language was particularly involved or clumsy. The paragraphs would have been written much more easily had certain terms been used. For example, *oxidation* is a single word that lets us refer to an atom losing one or more electrons. The gain by an atom of one or more electrons can be said by the one word, *reduction*. Similarly, an atom that yields electrons to another particle is a *reducing* agent and the particle that strips electrons from another particle is an *oxidizing* agent. *Electronegativity* is a single word that refers to the property of a particle to attract electrons from another particle and *oxidation potential* refers to the ease with which a particle gives up electrons. The metal that loses electrons is the *anode* and the metal whose ions are gaining electrons is the *cathode*.

Recall from the above paragraphs such a sentence as: But suppose we arrange a system in which one metal can lose electrons only if ions of another metal will take them on. By using the vocabulary that has been developed for use in such situations, the sentence becomes: But suppose we arrange a system in which one metal can be oxidized only if another metal is reduced. Do you see some of the reasons chemists tend to develop a more or less specialized vocabulary? Practice translating the clumsily written sentences into the straightforward language permitted by using the vocabulary that goes with the topic of oxidation-reduction.

12-2. Predicting the Voltage of Various Combinations of Half Cell Reactions. Having recorded the readings of your voltmeter in one or more electrochemical cells prepared in connection with

Instructional Activity 12-1, arrange with your neighbors to set up cells consisting of three or more half cell reactions. For example, what should a voltmeter read if you add to the original silver-copper cell an additional salt bridge connecting to a zinc half cell? Make your predictions "sight unseen" before actually hooking up the voltmeter. Can you select the combination of half cells you would need to arrange to produce a specified voltage? For example, can you arrange a complex cell that will cause the voltmeter to show precisely 2.0 volts?

12-3. Using "Commercial" Electrochemical Cells to Manage or Control Chemical Reactions. Here is an interesting hypothesis. If chemical reactions can produce electrical currents, then electrical currents can be used to produce chemical reactions. To explore this hypothesis, obtain several dry cells or a power converter whereby the usual 110-volt, Alternating Current, electricity of wall outlets can be converted to Direct Current and stepped down in voltage to only a few volts. This will be your "commercial" source of electrical current.

Place approximately 100 ml of 1.0 M copper sulfate in each of four 150 ml beakers. In one beaker, hang a copper-coated and a shiny iron, chromium-plated, or aluminum nail. In a second beaker, omit the copper-coated nail but include a shiny iron or plated nail. In the third beaker, hang a shiny nail attached to the positive terminal of a dry cell and insert a carbon electrode that is attached to the negative terminal of the dry cell. In the fourth beaker, hang a shiny nail that is connected to the negative terminal of a dry cell and insert a carbon electrode that is attached to the positive terminal of the dry cell.

Allow all beakers to stand for a few minutes and then observe the results. On the basis of this activity, what is one way in which an electrical current can be used to manage or control chemical reactions?

Fill two U-tubes with 1.0 M sodium iodide. Add approximately 1 ml of trichlorotrifluorethane and several drops of phenolphthalein solution to each U-tube. Shake thoroughly and observe. Attach the U-tubes to a ring stand or other suitable support. Use carbon electrodes to pass current from dry cells through both U-tubes but have the current enter the U-tubes from opposite directions. As soon as colors appear in the solutions in the U-tubes, remove the electrodes, shake thoroughly, and observe. What is a second way chemical reactions can be managed or controlled by using electrical current?

Arrange to have current from a dry cell pass through a solution of barium hydroxide (carbon electrodes) and a flashlight bulb. Prepare the barium hydroxide by adding an excess of the solid and then filtering in order to obtain a clear, saturated solution. While the current is flowing and the light bulb is glowing, add a drop at a time, 1.0 M sulfuric acid until the light is extinguished. Then add 10 more drops. Discuss your observations from the general idea of using electrical currents to manage or control chemical reactions. You might want to add a few drops of phenolphthalein to the barium hydroxide solution before adding the acid.

INSTRUCTIONAL ACTIVITIES: TEACHER

12-T1. Interactions Between Electrical and Chemical Phenomena. A demonstration that confronts students with several very enlightening observations involves the chemical reactions applied in the lead storage battery. Place two highly polished, preweighed lead strips in a battery jar containing 5 M sulfuric acid and wire into the circuit a voltmeter. Duplicate this set-up but include a power converter. With a minimum of discussion, perform the following procedures. Some teachers prefer to warn their students that a "quiz" may follow the observation phase of this type of presentation, others do not.

1) Invite a student to observe closely and describe the condition of all four lead strips.

2) Send current through the second set-up adjusting the voltage to approximately 5 volts. Allow the current to flow for 10 minutes or so.

3) Again, invite a student to observe and describe all four lead strips.

4) Shut off the current.

5) Distribute a quiz sheet based on confirming observations made during the demonstration by such a scheme as:

1. An accurate statement of something observed.
2. In direct conflict with the observations.
3. Goes beyond actual observations.

4. Was neither supported nor contradicted by any observation. Sample observations might be:

a. The voltmeter read the same both before and after the power source was disconnected.

b. Had the battery been charged at a higher voltage, the voltmeter would have shown a higher reading when the power source was disconnected.

c. The same chemical reaction occurred on each of the four lead strips.

d. For the set-up through which current was passed, both lead strips were the scene of the same chemical reaction.

e. Electrical current could be obtained from the newly charged lead storage battery at the same amperage and for the same length of time as delivered by the power converter.

f. If the newly charged lead storage battery was completely discharged, the lead strips would be returned to the condition they were in before the cell was charged.

g. After the cell was charged, the two lead strips would not have gained or lost the same amount of mass.

Inasmuch as this type of quiz-centered instructional activity is more instructional than grade determining, students may be invited to "mark their own papers" by having the instructor do whatever must be done to reveal the "correct" answers. Some items, of course, will demand discussion, especially those which might be claimed to go beyond actual observations or were neither supported nor contradicted.

12-T2. Faraday's Electrochemical Laws. In this exercise, a Daniell cell will be used to generate electrical current which, in turn, will be used to transfer lead from one to another lead strip in a "storage" cell. By including a milliammeter and voltmeter in the circuit, the number of coulombs can be equated with the grams of lead transferred and, thence, the number of electrons passed per lead ion transported can be calculated.

Prepare a Daniell cell by using well polished copper and zinc electrodes in 1.0 M copper and zinc sulfate solutions. When connecting these electrodes to the voltmeter and the milliammeter, as well as the two lead electrodes, be sure all connections introduce a minimum of resistance. Place the freshly polished, preweighed lead strips in a 250 ml beaker or battery jar containing 200 ml of 0.5 M lead nitrate. The new lead plating will form on the strip connected to the zinc electrode of the Daniell cell and a better coating will be obtained if this lead strip is placed in a porous cup inside and filled with the lead nitrate solution.

After a 45 minute or so current, remove the lead strips being careful to joggle loose a minimum of the transferred lead, rinse gently, dry, and weigh. For the sake of timing, you may prefer to set up the equipment and start the current 40 or 50 minutes before class time. This will allow the major portion of the class session to be devoted to the accompanying calculations and the discussion of results.

If the milliammeter shows an average current of 140 milliamps for 60 minutes, 504 coulombs would be involved. Assuming 1.60×10^{-19} coulomb per electron, this yields 3.15×10^{21} electrons or, 5.23×10^{-3} mole of electrons. If a weight transfer of 0.50 gram of lead accompanied the passage of the 504 coulombs, calculations show that approximately one atom of lead was transferred by the passage of each 2 electrons.

Discussion of the results from this exercise should insure that each student senses the interdependence between electron transfer and chemical reactions—an interdependence supposedly clarified vividly by the quantitative relationships brought out by this demonstration. But there is always the hazard that many of our students bog down in the calculations or devote all available time to carrying out the computations. Teachers learn to divide the available time so that the faster moving students gain the satisfaction from "discovering" quantitative relationships by doing their own computations but insuring that, before the dismissal bell sounds, all students will have seen where the computations were supposed to lead.

Audio-Visual Aids

12-AV1. "Development of Electrochemistry." Color. 19 min. Statens Filmsentral, Norway, for OCED in Association with the International Council for Education Films. Progressive development of electrochemistry. Catalysis. Technological applications.

12-AV2. "The Electric Cell." 50 frames. Jam Handy. Primary and secondary cells.

12-AV3. "Electrochemical and Electrolytic Cells." 44 frames. FOM. Oxidation-reduction reactions in test tubes and electrolytic cells with representative calculations.

12-AV4. "Electrochemical Cells." 15 min. Chem Study. Silver-copper cell, migration of ions, electron flow, and effects of changes in concentration.

12-AV5. "Electrochemical Reactions." Color. 13 min. Coronet. Energy conversions, ionization.

12-AV6. "Electrochemistry." 11 min. Britannica. Nature and applications of.

12-AV7. "Electrochemistry—Linking of Two Sciences." 40 frames. FOM. Dissociation and ionic conductivity. Conversion of chemical to electrical energy. Galvanic versus electrolytic cells.

12-AV8. "Electromotive Force." Color. 12 min. McGraw-Hill. Construction of E.M.F. series and relationships between E.M.F. and cell voltage and chemical activity.

12-AV9. "Faraday's Laws." Color. 15 min. McGraw-Hill. Demonstrates two laws. Animation to show mechanism of electrolysis.

12-AV10. "The Nature of Corrosion Anode and Cathode Processes." Color. 20 min. Rothacker. Electrochemical nature of corrosion, oxygen's role in.

Ideas for Additional Instructional Activities

12-AD1. Alexander, Classen, *Quantitative Analysis by Electrolysis.* New York: J. Wiley, 1919. Pages 73–78. Interesting from an historical point of view to see how analyses were carried out during another day. Procedures for copper, cobalt, and nickel.

12-AD2. Alyea, Hubert N., and F. B. Dutton, *Tested Demonstrations in Chemistry,* Easton, Pennsylvania: *Journal of Chemical Education,* 1960. Pages 60, 62, 73, etc. Reactions at the electrodes when electrical current is sent through a solution of potassium iodide, starch, and phenolphthalein and other demonstrations involving electrochemistry.

12-AD3. Alyea, Hubert N., "TOPS in Chemistry," *The Science Teacher,* 29: 10–17. May, 1962. Devices and procedures for projecting electrochemical phenomena.

12-AD4. Staff, "Avogadro's Number," *Science and Mathematics Weekly,* 2: 334. 1962. Electrolysis of copper(II) sulfate.

12-AD5. Bachman, G. B., and J. K. Farrell, "Special Exercises for Students in General Chemistry: II. Electrochemistry," *Journal of Chemical Education,* 10: 689. 1933. A dichromate cell to yield electrodeposition and electrochemical equivalent data.

12-AD6. Conway, B. E., and Mark Salomon, "Electrochemistry: Its Role in Teaching Physical Chemistry." *Journal of Chemical Education,* 44: 554–562, October, 1967.

12-AD7. Dillard, Clyde R., and Patty Hall Kammeyer, "An Experiment with Galvanic Cells for the General Chemistry Laboratory," *Journal of Chemical Education,* 40: 363–365. July, 1963.

12-AD8. Ehrenreich, Henry, "The Electrical Properties of Matter," *Scientific American,* 217: 195–204. September, 1967. As viewed from the perspective of quantum mechanics.

12-AD9. Ellis, R. H., and R. B. Rauch, "Determining Avogadro's Number," *Journal of Chemical Education,* 30: 460–461. 1953. Based on application of Faraday's Laws to quantitative electrolysis of dilute sulfuric acid.

12-AD10. Feldman, L., and A. T. Kuhn, "A Versatile Cell for Electrochemical Measurements," *Journal of Chemical Education,* 41: 390. July, 1964. Multicompartment cell for investigating electrode potentials, corrosion, and electro-organic reactions.

12-AD11. Fowles, G., *Lecture Experiments in Chemistry,* London: Bell and Sons, 1959. Pages 426–432. Comparison of the effects of electric charge on solutions of various substances.

12-AD12. Lauren, Paul M., "The Dynamics of an Electrolysis," *The Science Teacher,* 32: 60. October, 1965. Redox phenomena related to electrochemical decomposition using saturated sodium chloride solution.

12-AD13. Neidig, H. A., *et al,* "Electrical Conductance and Interaction of Ionic Solutions," *Chemistry,* 38: 25–31. October, 1965. Effects of ionic size, concentration, and temperature.

12-AD14. Parish, C., "The Number of Atoms in a Molecule of Hydrogen," In *Scientific Experiments in Chemistry,* New York: Holt, Rinehart, and Winston, 1959. Page 5. Based on the electrolysis of water.

12-AD15. Sanderson, R. T., "On the Significance of Electrode Potentials," *Journal of Chemical Education,* 43: 584–586. November, 1966. Quantitative relationships involved in electrodes and cell potential.

12-AD16. Stong, C. L., "How to Make an Electrochemical Cell," *Scientific American,* 217: 131–137. November, 1967. Volta's pile, historical, and fuel cell, modern.

12-AD17. Williams, L. Pearce, "Humphrey Davy," *Scientific American,* 202: 106–113. June, 1960. Widely known as the inventor of the safety lamp, also founded modern electrochemistry, tutored Faraday, and discovered sodium, potassium, barium, magnesium, strontium, and chlorine.

13

BONDS WITHIN MOLECULES

Although such observations are based on highly subjective evidence, students approaching the topic of chemical bonding begin to feel the real bite, the brain stressing crunch of chemistry. At one moment, we urge our students to parley their observations of a chemical event into a mental model that will apply to a whole universe of events. But we caution them sharply to not lose sight of what a mental model really is. Moments later, we catch ourselves using what is basically a mental model to construct an "explanation" for the properties of a molecule or the mechanism of a reaction.

There are types of students who sense extreme discomfort when they are subjected to intellectual gymnastics, circularity in reasoning, for example—gymnastics which don't seem quite right even though they can't word what it is that persists in cluttering up their thinking processes. If it is true that the principles and concepts especially involved in bonding theory mark a critical point along our students' pathways to success in their early encounters with chemistry, it follows that we must bend every effort to find effective instructional procedures and activities. And some of this effort may find us exploring and experimenting with strategies and techniques that depart from ways of teaching highly endorsed by time and tradition—ways of teaching focused directly, we hope, on the mental blocks or frustrating fuzziness quite symptomatic of impending diminishing enthusiasm for continued experiences in chemistry.

INSTRUCTIONAL ACTIVITIES: STUDENT

13-1. A Lesson in Words and What They Can and Can't Do for Us. To think about whatever it is that "glues" atoms together to form molecules—whatever are the forces which must be overcome before existent molecules can be broken apart or relied upon to reassemble the fragments into new molecules, is to bring our brains to bear on some of the most challenging aspects of chemistry. Chemists and chemistry teachers, especially those who write textbooks, are anxious to help you meet this challenge and have gone to great effort to pass along to you their ideas of what goes on in that invisible world within atoms and molecules. This they can do only by words—words that can't possibly do what they are supposed to do unless they mean to you what they mean to everyone else.

When it comes to using words to convey meaning, that is, to define what it is we are talking about, troubles are likely to arise for reasons which are as much the author's as the reader's fault. If an author defines something in terms of what the something *does* when we thought he was telling us what the something *is,* he may be guilty of misleading us. For example, if we want to know what an acid *is,* we may be confused if we get the definition: An acid is something that turns blue litmus paper red. Similarly, if we see a man using something labeled an acid to wash away the mortar spilled from between the bricks in a newly laid brick wall and ask him what an acid is, we might be confused more than enlightened were he to say, "An acid is a proton donor."

Operational definitions, those which state the properties of the thing being defined, are useful in certain situations. So are conceptual definitions, those which attempt to explain or account for the thing being defined. Less useful are definitions which simply say what the thing being defined is not. Nor are definitions useful which simply double back and restate what is being defined. Poor definitions also tend to use words that are less well known than the word being defined.

For this instructional activity, you are to examine definitions of some of the terms that are used in describing the forces which cause atoms to stick together or bond in molecules. Use the KEY to contrast these definitions with the meaning they carry as used by the authors of your textbook. For each definition, decide which of the statements applies best to the definition.

KEY:

1. A reasonably good *operational* definition, that is, it reveals the important properties of the thing being defined.
2. A reasonably good *conceptual* definition, that is, it provides a meaningful explanation of the thing defined.
3. A negative definition simply saying what the thing being defined is not.
4. The definition repeats itself.
5. The definition uses too many words which, in themselves, have little meaning.

a. Electronegativity—The opposite of oxidation potential.

b. Ion—An electrically charged particle containing more or fewer electrons than protons.

c. Ionization energy—The energy needed to convert atoms or parts of molecules to ions.

d. Ionic bonds—The electrostatic forces acting between anions and cations.

e. Valence—The combining capacity of an atom or ion.

f. Valence—The number of electrons needed to be gained or lost by an atom to achieve the electron configuration of the nearest inert gas on the periodic table.

g. Covalent—Mutual valency.

h. Chemical bond—The electrostatic attraction between the nuclei of two atoms and one or more of each other's electrons.

i. Ionic bond—A kind of bond which breaks when the substance goes into a water solution to produce a system which conducts electrical current.

j. Electropositive—Tendency to give up electrons.

k. Sigma bond—A bond formed by electrons in orbitals which seem to be blends of the usual or predominant orbitals in atomic architecture.

INSTRUCTIONAL ACTIVITIES: TEACHER

13-T1. Observations Versus Deductions Versus Mental Models or Theories. By the very nature of the topic, the study of the bonds within molecules requires our students to handle several types of knowledge simultaneously. We cite the properties of selected compounds in order to bring out orderly relationships or to identify puzzling discrepancies. Once an orderly relationship is identified we tend to deduce therefrom a state of affairs we hope applies to situations which go beyond our immediate observations. And when a system of deductions seems to offer the promise of being interrelated, we tend to build a mental model or theory that encompasses all of our observations and deductions. Once we construct a mental model and especially so if the model is quite elegant and logically consistent, we find ourselves turning to the model for explanations of the very observations and deductions from which the model was fabricated.

But there seems to be a valid place for this kind of mental gymnastics in the pursuit of science, especially chemistry. The important thing seems to be that our students recognize each type of knowledge for what it is and what it can do. It is this goal that is sought in the following instructional activity. The activity features a guide sheet that is intended to focus student attention during a homework assignment or class discussion. Although the guide sheet takes the form of a quiz, most students will need a clarification of the items which make up the key list.

Directions to the Student. For this activity, choose from the KEY list the type of knowledge represented by each of the statements which follow:

KEY:

1. An observation, that is, something actually observable.
2. A statement derived or deduced from one or more valid observations.
3. A piece or segment of a mental model.
4. A statement derived from a mental model or theory.
5. A false statement or unjustified deduction.

a. The nucleus of an atom is a point in space carrying a positive charge with the magnitude of the charge directly proportional to the number of protons making up the nucleus.

b. An electron is a negatively charged cloud with the dimensions of the cloud determined by the movement of the electron.

c. When each of two atoms have two electrons occupying or sharing the same orbitals, the atoms are joined by a double covalent bond.

d. If several different kinds of molecules all contain the same number of electrons and protons, all substances composed of these molecules have the same chemical properties.

e. The properties of substances are influenced both by the numbers and arrangement of the atoms making up the molecules of the substances.

f. If a solution conducts the electrical current then the solute's molecules contained ionic bonds.

g. The kinds and strengths of bonds or forces between molecules are likely to be determined by the kinds and strengths of bonds or forces between the atoms within the molecules.

h. The combining capacity of an atom is determined by the number of electrons versus the number of unfilled or partially filled orbitals in the outermost energy levels.

i. The atoms which make up molecules of carbon monoxide have equal combining capacities.

Audio-Visual Aids

13-AV1. "Coordinate Covalent Bond." 40 frames. FOM. Coordinate covalent bonds in polyatomic ions, certain molecules, and Werner complexes.

13-AV2. "The Covalent Bond: Molecular Orbitals." 44 frames. FOM. Molecular and hybridized orbitals to describe the structure of complex molecules.

13-AV3. "Covalent Bonds—Covalent Structures." 44 frames. FOM. Uses a variety of models to relate electronegativity to types of covalent bonds and structures.

13-AV4. "The Ionic Bond." 44 frames. FOM. The disperse nature of ionic bonds, the significance of empirical formula, unit cells and lattice structures, and related radius ratios.

13-AV5. "Macromolecular Bonds." 40 frames. FOM. Fundamental characteristics of such network solids as diamond, silicon dioxide, and graphite. Relies on concept of strong covalent bonds in a three dimensional aggregate.

Ideas for Additional Instructional Activities

13-AD1. Dickieson, Samuel C., "Diagramming Ionic and Covalent Compounds," *The Science Teacher,* 26: 123–124. March, 1959.

13-AD2. Howald, Reed A., "Bond Energies in the Interpretation of Descriptive Chemistry," *Journal of Chemical Education,* 45: 163–172. March, 1968. Bonding, valences, and stability.

13-AD3. King, L. Carroll, "Molecular Architecture," *The Science Teacher,* 33: 27–34. September, 1966. History. Types of structures.

13-AD4. Nechamkin, Howard, "Demonstrating Relative Covalent-Ionic Nature in Salts," *The Science Teacher,* 31: 58–59. March, 1964.

13-AD5. Sanderson, R. T., "Principles of Chemical Bonding," *Journal of Chemical Education,* 38: 382–391. August, 1961. Characteristics and conditions which lead to various types of bonds.

13-AD6. Sanderson, R. T., "The Nature of 'Ionic' Solids: The Coordinated Polymeric Model," *Journal of Chemical Education,* 44: 516–523. September, 1967. Includes discussion of non-molecular solids, electron density, and ionic radii.

13-AD7. Veguilla-Berdecia, Luis Antonio, "The Nature of the Chemical Bond," *The Science Teacher,* 36: 74–79. February, 1969. As seen through a study of the hydrogen molecule.

13-AD8. Wilmshurst, J. K., "Ionic Character, Polarity, and Electronegativity," *Journal of Chemical Education,* 39: 132. March, 1962. Includes summary.

14

BONDS BETWEEN MOLECULES

Knowledge concerning the bonds which hold atoms together in molecules is obtained pretty much by investigating the events and circumstances involved when these bonds are broken or reformed. In similar fashion, knowledge concerning bonds between molecules is based on observations made when these bonds are stretched, broken, or reformed. Whereas bond breaking events *within* molecules are very likely to involve truly chemical reactions, only physical changes are involved when inter-molecular bonds are tampered with. And changes in state, solid, liquid, or gas, are the most obvious physical changes. Although our students may realize that the physical state of a substance is closely tied to temperature, they may be less accustomed to looking upon changes in state as being bond stretching, breaking, or reforming processes.

Compared to intra-molecular, the phenomena accompanying inter-molecular bonds are more easily observed. But they are no more easily traced to clear-cut mental models of atomic and molecular structure. In fact, we turn increasingly to the internal structure of molecules for theoretical explanations of the external "stickiness" or bonding tendency of molecules. And again, some of our students may remind us of their uneasiness with what they interpret to be circular logic—an uneasiness that can be avoided somewhat by playing down hard, cold cause-effect connotations and emphasizing sheer concomitance.

In general, our students are more comfortable when we give them clear-cut cause-effect type explanations. They feel that something is missing from explanations which state the probability that such and such an event will precede, accompany, or follow the occurrence of the phenomenon being "explained." But the closer we take our students to the interface between what is known for sure and what can only be surmised about chemical phenomena, especially the phenomena involved in bonding, the more we find ourselves having to settle for increasingly relativistic explanations.

INSTRUCTIONAL ACTIVITIES: STUDENT

14-1. Relating the "Stickiness" Between Molecules to Other Properties of Substances. The following table shows some of the properties of four substances. Figure 14-1 suggests the probable

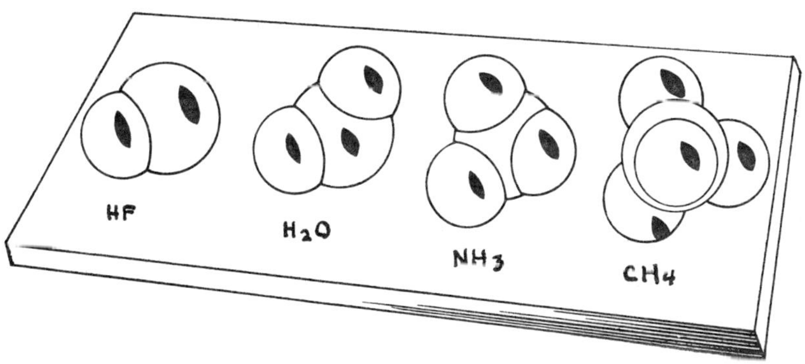

Figure 14-1

structure of the molecules of these substances. Although each of the four substances has a unique number of atoms in its molecules, note that all four of the molecules consist of the same number of protons and electrons. Also, note that these four substances were selected because they differ markedly in boiling points.

For this instructional activity, keep in mind that we are concerned with the nature of the bonds between molecules—that we are looking for the sources of the forces which cause molecules to stick to each other. An explanation of these forces is likely to be put together from several different kinds of knowledge. Use the following definitions as a KEY list to see if you can recognize each type of knowledge as illustrated in the statements which follow.

1. Observation—A fact or occurrence duly noted and, usually measured or otherwise described quantitatively.
2. Assumption—Taking for granted that something will be such and so simply because it is a part of a well known and widely accepted state of affairs.

KEY:

3. Deduction—A tentative conclusion based on a combination of observations, assumptions, or inferences.
4. Hypothesis—A tentative explanation so stated as to await or invite proof or disproof of its validity.
5. Mental Model—A comparison intended to help visualize in a simplified way something that cannot be observed directly.

The statements which follow involve the properties of four substances and their apparent relationships to the kinds of bonds which cause their molecules to stick or cling together.

Substance	Formula	Molecular Mass	Boiling Point
Methane	CH_4	16.03	−161.5°C.
Ammonia	NH_3	17.03	− 33.4
Water	H_2O	18.02	100.0
Hydrogen Fluoride	HF	20.01	19.9

a. The higher the velocity with which two particles are moving, the greater the probability that they will break apart after colliding.

b. The farther apart two particles are, the less effective are the attractional or repelling forces between them.

c. The stickiness between molecules originates within the molecules.

d. The strength of the bonds between molecules is not solely determined by the numbers of oppositely charged particles in the molecules.

e. The fewer atom-to-atom bonds within a molecule, the greater the strength of the molecule's molecule-to-molecule bonds.

f. The temperature at which a substance boils provides a measure of the stretchability and, hence, breaking strength of the bonds between the molecules of the substance.

INSTRUCTIONAL ACTIVITIES: TEACHER

14-T1. Confronting Students with Observations Involving Selected Bond Forming, Stretching, or Breaking Phenomena. Establish and, possibly, exhibit on the chalkboard a set of criteria which can be used to categorize types of bonds. Include criteria whereby inter- can be distinguished from intra-molecular bonds and highly ionic from highly covalent. Depending upon how far the class has advanced in its study of bonding phenomena, you may choose to include criteria for identifying van der Waal's forces, hydrogen bonding, molecular crystal, network solid, etc. Sample criteria follow.

Tampering with inter-molecular bonds is likely to affect the physical properties of a substance whereas stretching or breaking intra-molecular bonds leads to the formation of substances with different chemical properties.

The greater the separation of two elements on the electronegativity scale, the greater is the ionic character of bond between them.

The greater the ionic character of intra-molecular bonds, the higher is the electrical conductivity of the substance when melted or of its water solution; conversely for covalency.

Intra-molecular bonds involving atoms of the same element are purely covalent. In general, the more two atoms differ in nuclear charge, the less covalent is the bond between them.

Hydrogen bonds, usually weaker than ionic or covalent bonds, occur when a highly electronegative atom's bond to a hydrogen atom upsets the charge symmetry of the hydrogen atom thus allowing the hydrogen atom to form additional bonding.

Van der Waal's forces become the dominant source of inter-molecular stickiness when each atom within the molecule has the electronic structure of an inert-gas atom and, hence, electrostatic attractions between molecules arise solely from attractions between the nuclei of atoms and nearby electron clouds of neighboring atoms.

Having established a set of criteria whereby the nature of specific bond affecting events may be discussed, confront your students with a set of events and invite efforts to categorize each event in terms of the kinds of bonds being formed, stretched, or broken. Sample events follow. Discussion may follow each step.

a. Heat a sample of paradichlorobenzene until it melts. Continue heating and collect the vapor on the bottom of a watch glass. Pour some of the melted paradichlorobenzene onto an asbestos mat. After discussing the "freezing" of this sample, ignite it with a burner. CAUTION: Ignite the vapor from a small sample being heated in a hard glass test tube. Try to pass electrical current through (1) solid and (2) fused samples of paradichlorobenzene. Try to dissolve a sample in water and test the conductivity of the "solution." Place a large crystal on a hard surface and shatter it with a sharp hammer blow.

b. Freeze distilled water and substitute the ice for the paradichlorobenzene in the previous activity. For the "solution" step, use ethanol for the solvent. You may also choose to use an immiscible "solvent."

c. Crush a large chunk of rhombic sulfur. Try to dissolve a sample in water. Dissolve a sample in carbon disulfide. Evaporate a sample of the solution. Try to pass electrical current through (1) a chunk of rhombic sulfur, (2) a pile of pulverized sulfur, (3) a water "solution," and (4) a carbon disulfide solution of sulfur. Heat a sample of pulverized rhombic sulfur until it melts. Pour some of this straw-colored liquid into a filter paper in a funnel and allow the liquid to freeze. Continue heating the residue until it appears as a dark red, highly viscous liquid. Pour some of this liquid into ice-water. Burn small samples of the (1) rhombic, (2) monoclinic, and (3) amorphous sulfur.

d. Substitute sodium iodide for the paradichlorobenzene in "a" but omit the steps which are compromised by the high boiling point of sodium iodide. *Use caution* when attempting to pass electrical current through fused sodium salts and especially so when cleaning up the residue lest metallic sodium has been produced.

Audio-Visual Aids

14-AV1. "Structure of Water." Color. 14 min. McGraw-Hill. Properties of water traced to inter- and intra-molecular bonds.

Ideas for Additional Instructional Activities

14-AD1. Alexander, Leroy E., "X-Rays, Crystals, and Life," *Chemistry,* Part I, 37: 6–11, August, 1964. Part II, 37: 5–7, September, 1964.

14-AD2. Alyea, Hubert N., and F. B. Dutton, *Tested Demonstrations in Chemistry,* Easton, Pennsylvania: *Journal of Chemical Education,* 1960. Page 25. Floating one-quarter inch diameter paper discs that are black on one side and white on the other in an Erlenmeyer flask containing approximately 50 ml benzene and 50 ml water produces an interesting observation. Use twenty or so discs and compare results with those obtained when the liquid system consists of 50 ml carbon tetrachloride and 50 ml water. (May substitute TTE.)

14-AD3. Choppin, Gregory R., "Water — H_2O or $H_{180}O_{90}$?," *Chemistry,* 38: 6–11. March, 1965. Models which have been proposed for the structure of liquid water.

14-AD4. Clever, H. Lawrence, "The Hydrated Hydronium Ion," *Journal of Chemical Education,* 40: 637–640. December, 1963. How the observed properties of aqueous solutions are translated into a model of unobservable events and circumstances.

14-AD5. Derjaguin, Boris V., "The Force Between Molecules," *Scientific American,* 203: 47–53. July, 1960. Measuring intermolecular forces and demonstrating their electromagnetic origin.

14-AD6. Zeldin, Martel, "An Introduction to Molecular Symmetry and Symmetry Point Groups," *Journal of Chemical Education,* 43: 17–26. January, 1966. Proper rotation axis C_n, plane of symmetry, center of symmetry, etc.

14-AD7. Lonsdale, Kathleen, "Disorder in Solids," *Chemistry,* 38: 14–19. December, 1965. The limitations of crystal models, thermal vibrations, structural disorder.

14-AD8. Pearson, A. David, "Laser Beams—How They Are Generated," *Chemistry,* 38: 8–11. April, 1965. Basic principles involved in generating laser beams in solid crystals.

14-AD9. Peiser, H. Steffen, and John L. Torgesen, "Crystal Growth as Chemical Research," *Chemistry,* 38: 14–20. September, 1965. How crystals are related to mineralogy, metallurgy, and organic structures. Ideas for growing crystals.

15

METALS AND THE METALLIC BOND

Our students are quite familiar with the hardness, lustre, electrical conductivity, and springiness of metals. Their world is filled with tools and devices which bear witness to how readily metals can be hammered into sheets, drawn into thick or thin wires, and deformed into highly intricate shapes. Burned fingers may have taught them the relatively low specific heat of metals and the efficiency with which metals conduct heat. They may be surprised to learn that some metals have melting points well below ordinary room temperatures but the tendency for metals to remain solid at relatively high temperatures is general knowledge.

The properties of metals can be traced to the peculiar manner in which their atoms bond. Herein lies one of the valuable features of studying the topic of metals and the metallic bond. Because our students are familiar with the observable properties of the metals, it is easier to take them into the unobservable realm of individual atom-to-atom interactions and have them come out with ideas which enrich their comprehension and appreciation of what they can actually observe.

INSTRUCTIONAL ACTIVITIES: STUDENT

15-1. Metals Versus Nonmetals. Think about the following statement: Of all the elements, about two-thirds are metals, about one-sixth are nonmetals, and about one-sixth have properties intermediate between those of metals and nonmetals. Look at a periodic table of the elements and decide what features of the structures of the atoms of the elements have the most to do with identifying an element as a metal versus a nonmetal. Hint—highly nonmetallic elements are in the northeast and highly metallic elements are scattered throughout the left hand portion of the periodic table. And don't overlook the "see below" notices in the areas of elements 57 and 89.

15-2. Winning Metals from Their Ores. Very early in the evolving human drama man realized that he could do things to the stones and rocks in his environment that would make of them better materials from which to fashion his tools and weapons. As you know, the Stone Age was followed by Ages of Bronze, Iron, and Steel. Today, it is difficult to choose a technological label that characterizes man's reliance upon any one metal. And although we might say that this is the Age of Alloys, metals formed by blending carefully determined amounts of two or more metallic elements, people might argue that man-made materials, especially plastics, characterize more faithfully today's technological world.

In any case, the extraction of metals from their ores continues to be an important part of applied chemistry. Furthermore, the chemistry of metallurgy provides a particularly inviting opportunity for us to visit the world of atoms, crystals, ions, and molecules and to see how their reactions can be managed or controlled. It is also interesting to realize that to extract a metal from its ore, the chemist may very well reverse reactions which occurred far back in the history of the earth. In this respect, the chemist adds to our knowledge of the events and circumstances which may have accompanied the origin of the earth and its adolescent years.

The extraction of lead from lead ores was one of man's earliest metallurgical successes. One lead ore is galena, PbS. This mineral can be dug from the earth's crust in seams or veins where it has been concentrated by the earth's chemistry processes. You can begin this activity with a chunk of galena. Or you may prefer to roll back the ages of geological time and begin the activity with a sample of lead and of sulfur not yet combined to form the lead sulfide. If so, shave the lead into thin strips and mix with sulfur in a crucible. Supposedly, the earth was very hot during its earliest years. Duplicate this state of affairs by heating your lead-sulfur mixture (*hood*) as hot as you can. You now obtain a product that could well have existed in the earth's rocks ever since the earth took its present form.

To retrieve the lead, you will need to not only raise its temperature again but also introduce another element more likely to combine with the sulfur. To solve these problems, shape a small cup in a charcoal block or briquette. Rely on the carbon of the charcoal to take care of the sulfur and raise the temperature of the galena by placing it in the cup-shaped depression and use a blowpipe to direct a hot flame against the ore. A small pool of molten lead will mark the success of your effort to win the lead from its ore.

You can visit more of the chemistry of metallurgy by way of winning copper from one of its ores, malachite, $CuCO_3 \cdot Cu(OH)_2$. Place a sample of this ore in a test tube and heat until no further changes occur. Now arrange to have a current of hydrogen gas flow over the hot copper oxide (Figure 15-1). As soon as the copper plates out on the walls of the test tube, dissolve the copper with a few mls of 6 M nitric acid. Use electrochemical techniques to recover the copper from this solution.

15-3. The Corrosion of Metals. At the check-out table you will find a collection of various kinds of nails from the local hardware. You can see that some of these nails are the ordinary iron or steel variety, some are especially hardened so they will drive into cement without bending, some are made of aluminum, and some are coated with zinc. The "coppery" colored nails may be made of copper or some kind of alloy. Take one of each kind of nail and devise a system for identifying each nail even if its external appearances may be changed drastically during this instructional activity.

Place each nail in a small test tube. Label each test tube and then add enough of one of the following reagents (assigned by your teacher) to cover the nail: NaOH, KOH, Na_3PO_4, $Na_2Cr_2O_7$, Na_2CO_3, $Na_2C_2O_4$, NaCl, KNO_3, NaSCN, HCl, HNO_3, H_2SO_4, distilled water. Each of these reagents, except the water, is approximately 1 M. Use the remainder of the laboratory period to prepare a table which will enable you to record the appearance of not only your set of nails but also all of your classmates' sets on the following day.

On the following day, visit each of your classmates' "exhibit" and record whatever evidence you can find that will help you decide whether or not the nail and the reagent have interacted. Where you

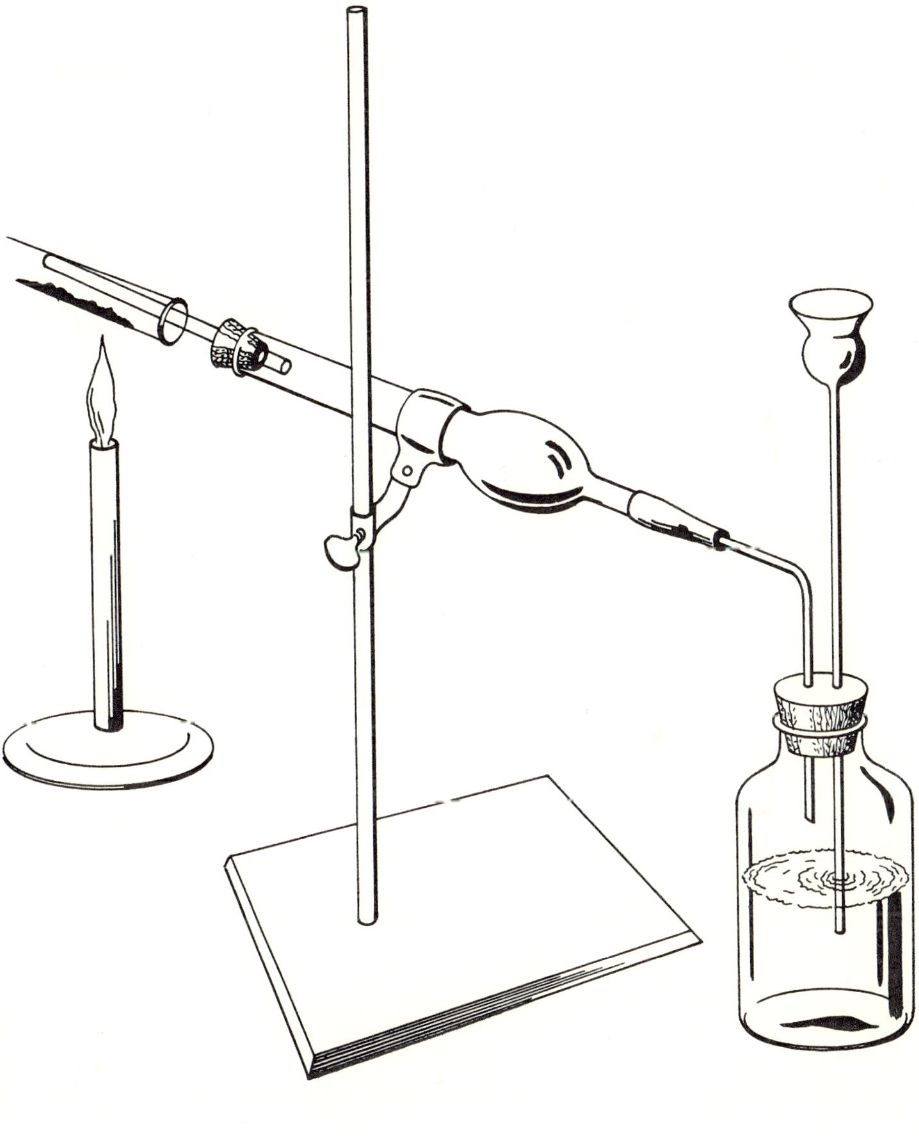

Figure 15-1

find evidence suggesting that a reaction has occurred, write the most probable equation for the reaction. In addition, make special note of any reactions which seem to be especially thought-provoking.

Looking ahead to further investigations of the corrosion of metals, it is possible to observe the effects of certain reagents on metals more effectively if the reagent is distributed through a semisolid medium such as gelatin. If you see where this technique can help you with an investigation, plan to add the reagent of your choice to an agar-agar gel. To prepare such a gel, heat about 200 ml of water to a gentle boil, remove from the heater and stir in 2 g of powdered agar-agar. Stir until the agar-agar is completely dispersed. While the gel is still hot and, hence, not yet gelled, pour off whatever volume you will need for your investigation and add a few milliliters of whatever reagent your investigation calls for.

Here are some examples of the kinds of investigations other students have thought of after studying their test tube sets of nails in various kinds of reagents. Place a straight nail and a bent nail in a petrie dish and cover both nails with agar containing KSCN solution. Wrap a copper wire around one nail, a strip of magnesium ribbon around a second nail, leave a third nail as it is and cover all three nails with agar containing KSCN. Arrange an iron, an aluminum, a zinc, and a "copper" nail in a petrie dish so they do not touch each other. In a second petrie dish, arrange a similar set of nails so that each nail touches two other nails. Compare the results after both sets of nails have been covered with agar containing one or another reagent. Does it make a difference which nail is touching which other nails? Compare the effects of various "gelled" reagents on ordinary versus "hardened" steel nails. Does the hardening process seem to be more a physical than a chemical process?

Audio-Visual Aids

15-AV1. "Crystals and Their Properties." 40 frames. FOM. Representative models to show properties of crystals classified according to bond types.

15-AV2. "Crystals, and Their Structure." 22 min. Modern Learning Aids. Fundamental crystallography.

15-AV3. "Metals and the Metallic Bond." 40 frames. FOM. Experiments to bring out the physical properties of metals. Develops the theory of the metallic bond, the mobility of the valence electrons, and the crystalline nature of metals. Touches on ionization energies and oxidation potentials.

15-AV4. "Packing of Atoms in Crystals." 32 frames. Britannica Filmstrips. Arrangements, role of imperfections.

15-AV5. "Shapes and Polarities of Molecules." Color. 21 min. Chems. Polarity as a function of molecular symmetry.

15-AV6. "Shapes and Properties of Molecules." 32 frames. Britannica Filmstrips. Boiling points, acid-base properties, crystal form, reaction rates, electronegativity as functions of molecular symmetry.

15-AV7. "A Study of Crystals." Color. 17 min. Journal Films. Laboratory demonstrations and animation techniques to show sublimation, crystallization from melt, and mixed fusion techniques.

Ideas for Additional Instructional Activities

15-AD1. Azaroff, L. V., *Introduction to Solids,* New York: McGraw-Hill, 1960. Structures of metals and characteristics of various types of packing.

15-AD2. Benfey, Theodor, "Geometry and Chemical Bonding," *Chemistry,* 40: 20–26. May, 1967. Reducing various types of bonds to a common set of forces.

15-AD3. Bent, Henry A., "The Tetrahedral Atom," Part I, *Chemistry,* 39: 9–13, December, 1966. Part II, 40: 8–15, January, 1967. The geometry of a tetrahedronal atom based on chemical evidence.

15-AD4. Carmody, Walter, and Jack Wiersma, "A Study of the Silver Tree Experiment," *Journal of Chemical Education,* 44: 417–420. July, 1967. Reactions between copper and silver.

15-AD5. Cottrell, A. H., "The Nature of Metals," *Scientific American,* 217: 90–100. September, 1967. From the perspective of the metallic bond.

15-AD6. Gehman, W. G., "Standard Ionic Crystal Structures," *Journal of Chemical Education,* 40: 54. 1963. Excellent figures and diagrams are used to discuss cubic and hexagonal closest-packing atom lattices.

15-AD7. Habashi, Fathi, "Kinetics of Corrosion of Metals," *Journal of Chemical Education,* 42: 318–322. June, 1965.

15-AD8. Hausman, Eugene A., "The Uses of Nobility," *Chemistry,* 37: 6–11. October, 1964. Occurrence, refining, properties, and uses of the platinum metals.

15-AD9. Pauling, Linus, *The Nature of the Chemical Bond,* Ithaca, New York: Cornell University Press, 1960. Page 393, third edition. The metallic bond, interatomic distances, lattice structures, electronic structures, intermetallic compounds, and other properties of metals.

15-AD10. Sime, R. J., "Some Models of Close Packing," *Journal of*

Chemical Education, 40: 61. 1963. Foam plastic models of unit cells, hexagonal close-packed, and face-centered cubic lattices.

15-AD11. Fernelius, W. C., and R. F. Robey, "The Nature of the Metallic State," In *Supplementary Reading for Chemical Bond Approach,* Easton, Pennsylvania: Chemical Education Publishing Co., 1960. Pages 125–140. Electron gas theories, solution of metals in nonmetallic solvents, crystal structures, alloys, and intermetallic compounds. Importance of metals as a topic in introductory chemistry course. See also, "An Introduction to the Electron Theory of Metals," by R. A. Lefever, pages 141–145 of the same book.

15-AD12. Swann, Peter R., "Stress-Corrosion Failure," *Scientific American,* 214: 73–81. February, 1966. Certain environments can cause metal alloys to give way unexpectedly.

15-AD13. Staff, "The Crystal Structure of Metallic Elements," *Science and Math Weekly,* 3: 58. 1962. Constructing models of body-centered, face-centered, hexagonal closest-packing, and icosahedral crystal lattices. See also page 94.

15-AD14. Verhoek, Frank H., "What Is a Metal?," *Chemistry,* 37: 6–11, November, 1964. Properties of metals and the forces which hold metallic atoms together.

15-AD15. Webb, Valerie J., "Hydrogen Bond 'Special Agent'," *Chemistry,* 41: 16–20. June, 1968. Properties of and evidence for.

16

QUANTUM THEORY

Only a few students enter the chemistry course already familiar with significant aspects of quantum theory. These students, having acquired knowledge of quantum theory through independent study or some fortuitous circumstance, find pride in discussing their interpretation of the topic. For the teacher who is preparing to introduce the topic, this can be both good and bad. Some students may have acquired a sense of foreboding mystery so likely to surround sophisticated topics and this sense could be deepened if their peers exaggerate the capabilities needed to comprehend quantum theory. Other students may be encouraged to know that their peers have already proved capable to handle the topic.

There is the challenge to present quantum theory to beginning chemistry students in a way that suggests its elegance and compromises neither its spirit nor function. Adding to this challenge is the necessity to fill in the gaps created where students' backgrounds in mathematics preclude retaining the quantitative logic that accompanied the development of quantum theory.

For the bulk of our students, our goal may be to help them see how quantum theory contributes to the total endeavor of describing chemical events. For sure, all students must sense the inevitable involvement of energy in chemical reactions. Equally axiomatic is the need to sense the particle nature of matter. It follows that to contemplate the particle nature of energy holds promise of integrating the matter and energy phases of chemical events. And this would appear to be good teaching.

INSTRUCTIONAL ACTIVITIES: STUDENT

16-1. Energy, Too, Comes in Particles. The story of chemistry is very much the story of particles. It was a giant step when men first saw in the dull blur of matter the fantastically tiny, totally invisible atoms and molecules of which all matter is composed. It was another master stroke of human intelligence when men visualized all atoms as being put together from only a few fundamental particles, electrons, protons, neutrons, and so forth.

There is equal elegance in the intellectual achievement whereby man has learned to visualize heat, light, electricity, in fact, all forms of energy as existing in equally fantastically tiny particles or bundles. Actually, mankind's development of atomic theory has gone hand in hand with his development of quantum theory, quantum being the name for an individual bundle of energy. In fact, the assumption that energy in matter is in quanta helped establish the structure of atoms.

The purpose of this instructional activity is to help you see how a major scientific concept is created. We want you to see that the quantum theory of the nature of energy and its interaction with theories regarding the structure of atoms coalesced from the individual efforts of men and women pursuing their own glimpses or ideas of things. Usually, these glimpses began with an observation of something or other. The observation, backed up by background and related information, led to an hypothesis or hypotheses. Hypotheses, in turn, lead to some kind of verification, usually by experimentation or using the hypothesis to predict how something should turn out. Hypotheses, once they are verified, are promoted to the status of scientific principles, concepts, or theories.

For this instructional activity, you are given the names of some of the men who have contributed to the development of the quantum theory and its interaction with the atomic theory. Each name is accompanied by either an observation, hypothesis, experiment, or conclusion involving one of their major contributions. Choose the scientist whose last initial is closest to yours and see if you can track down the other three kinds of information involved in his contribution.

a. Max Planck—A hot body cannot emit or absorb light of a given wave length in less than a minimum amount.

b. J. J. Thomson—The electrical conductivity of certain metals is influenced by shining light on the metals.

c. Albert Einstein—If the photoelectric effect depends upon dislodging electrons from metals, then each electron being dislodged has absorbed all of the energy in at least one photon.

d. Louis de Broglie—If energy travels in waves and if a moving electron is a bundle of energy, then individual electrons can be described by the same dimensions as are used to describe an energy wave.

e. W. Heisenberg, E. Schrodinger, and P. A. M. Dirac—If energy exists in individual bundles and if energy is emitted from a vibrating or oscillating source, then the vibrating source must be limited to only those frequencies which are permitted by the equation: Wave length equals the velocity of the wave divided by the frequency of the vibrating source.

f. Niels Bohr—In the hydrogen atom, the electron revolves in an orbit with radius 0.530 Å.

g. G. E. Uhlenbeck and S. A. Goudsmit—Each electron spins either clockwise or counterclockwise about an axis.

h. W. Pauli—Two electrons can occupy the same orbital only if their spins are opposed.

i. Niels Bohr—The hydrogen atom's electron can exist at ground state or at only certain excited or higher energy levels and when the electron changes from a higher to a lower energy state, the energy of the light quantum that is emitted is described by the equation: Planck's constant times frequency equals the difference between the higher and the lower energy levels.

16-2. The "Address" of Each Element's Electrons. Just as you have an address whereby you can tell other people where you live, chemists have devised a system whereby they can identify the orbital most likely to be occupied by each of the electrons built into any kind of atom. Your address may involve a state, city, street and number, and, perhaps, a specific apartment. To identify the "address" of an electron requires four quantum numbers; one for the principal energy level, a second for the sub-shell within this principal energy level, a third for the orientation in space, and a fourth to designate the direction of spin and thus distinguish the electron from its orbital mate, if it has one.

To acquaint you with the addresses of each element's electrons, particularly those most likely to engage in bondings with other atoms, list each of the elements in the order of increasing atomic number and complete the following table:

Atomic No. of Element	Address of the Electron Not Present in Atom of Next Lower Atomic No.	Addresses of the Atom's Valence Electrons
1. Hydrogen	ls	ls
2. Helium	ls + ½	—
30. Zinc	5s	$5s^2$

While completing this table, don't be carried away with the mechanics and forget what it is you are doing. Actually, to know the orbital that an electron is "at home" in, the number of other valence electrons it has as neighbors, and the orbital it is likely to share when bonding to another atom can be a big help in understanding how and why reactions occur.

INSTRUCTIONAL ACTIVITIES: TEACHER

16-T1. Formulas and Equations. Instructional time can often be invested in helping students grasp and appreciate how a formula or equation can be a highly informative description of a principle or concept. There is always the hazard that prior instruction has caused a fraction of our students to associate formulas with meaningless mathematical chores. Quantum theory, in contrast, provides a good opportunity to encourage our students to reexamine their habits and attitudes involving formulas and equations.

At a psychologically appropriate time following introduction of the fundamental ideas of quantum theory, confront your students with an array of quantum theory related equations. Include a legend showing what each symbol stands for. Follow the array and legend with a KEY list such as:

KEY:
1. An increase in the first factor will be accompanied by an increase in the second.
2. An increase in the first factor will be accompanied by a decrease in the second.
3. A change in the magnitude of the first factor will not, necessarily, be accompanied by a change in the magnitude of the second.

a. $\lambda = \frac{c}{\nu}$	λ—wave length.
b. $E = h\nu$	c—velocity of light. (3.0 $\times 10^{10}$ cm/sec.)
	ν—frequency.
c. $E_2 - E_1 = h\nu$	h—Planck's constant (9.52×10^{-14}kcal sec/mole).
d. $E_n = -2m\left(\frac{\pi Z e^2}{n\,h}\right)^2$	E_1 and E_2—Initial and final energy levels.
	E_n—total energy of an electron.
	m—mass of an electron.
	Z—Nuclear charge (atomic number).
	e—Charge on an electron (1.60×10^{-19}joule/volt).
	n—principal quantum number of the electron.

1. An increase in the wave length of a photon.—The velocity with which the photon travels.
2. The number of energy levels through which an electron "falls."—The wave length of the photon that is emitted.
3. The principal quantum number of an electron.—The energy represented by the electron.
4. The energy represented by a photon.—The frequency of vibration or oscillation of the photon's source.
5. The color of light emitted by a source.—The mass of the object whose vibrations are producing the light.
6. The atomic number of the atom with which an electron is associated.—The mass of the electron.
7. The energy represented by a photon.—The wave length of the photon.
8. The number of energy levels through which an electron "falls."—The energy represented by the photon that is emitted.

Ideas for Additional Instructional Activities

16-AD1. Alyea, Hubert N., "Photochemistry," *Journal of Chemical Education,* 44: A145. February, 1967.

16-AD2. Berry, Stephen R., "AC_3 Resource Paper: Atomic Orbitals," *Journal of Chemical Education,* 43: 283–296. June, 1966.

16-AD3. Bragg, Sir Lawrence, "X-Ray Crystallography," *Scientific American,* 219: 58–70. July, 1968. Past, present, and promise.

16-AD4. Eichinger, Jack W., "Anticipating 'Valences' from Electron Configurations," *Journal of Chemical Education,* 44: 689–690. November, 1967.

16-AD5. Ferreira, Ricardo, "Molecular Orbital Theory, an Introduction," *Chemistry,* 41: 8–15. June, 1968. Orbitals as wave functions.

16-AD6. Gamow, George, "The Exclusion Principle," *Scientific American,* 201: 74–86. July, 1959. Orbits, periodicity, quantum mechanics.

16-AD7. Garrett, Alfred B., "Quantum Theory: Max Planck," *Journal of Chemical Education,* 40: 262–263. 1963. Reasoning used by Planck in deriving his equation. Planck's constant in erg-sec.

16-AD8. Gentile, Philip S., "Solid-Solid Interactions," *Chemistry,* 39: 28–33. February, 1967. Contrasting with reactions occurring in the gaseous or liquid phases.

16-AD9. Hirschfelder, Joseph O., "A Forecast for Theoretical Chemistry," *Journal of Chemical Education,* 43: 457–463. September, 1966. History, present, and future of quantum mechanics.

16-AD10. Lambert, Frank L., "Atomic Orbitals from Wave Patterns," *Chemistry,* Part I, 41: 10–15, February, 1968; Part II, 41: 8–11, March, 1968. Visualizing three-dimensional waves.

16-AD11. Staff, "Molecular Orbital Theory," *Chemistry,* 40: 26–27. January, 1967. Based on molecules of chlorine and oxygen.

16-AD12. Stong, C. L., "The Amateur Scientist: How to Recreate the Apparatus with which the Charge of the Electron Was Measured," *Scientific American,* 200: 175–186. May, 1959.

16-AD13. Yamanashi, Bill S., and Andrew V. Nowako, "Recombination of Iodine via Flash Photolysis," *Journal of Chemical Education,* 45: 705–710. November, 1968. A chemical kinetics experiment in physical chemistry.

16-AD14. Williams, Buford, "Inexpensive Apparatus for Spectrophotometric Experiments," *Chemistry,* 41: 28–30. March, 1968.

17

REACTION MECHANISMS AND THE KINETICS OF REACTIONS

The topic of this chapter invites questions involving why, how, and how fast chemical reactions occur. It is the "why" kind of question that students seem to be most anxious to ask. Especially early in the course, there is a type of student who becomes sincerely irked when he doesn't receive a satisfactory answer to such questions as: "Yes, I know that the dense white precipitate is silver chloride but *why* should silver chloride be any more likely to precipitate out of the system than sodium chloride?"

At the other extreme, there are students who raise no questions at all and go about their laboratory exercises as though chemical reactions occur simply because the textbook, laboratory manual, or instructor has said they will. These students would be easier to teach if they would become motivated to ask any kind of question.

Sooner or later, each teacher should establish with his students his interpretation of the difference between "why" and "how" types of questions—difference more fundamental than purely semantic or "ivory tower" argument. The optimum goal seems to be to establish a communications environment that enables us to describe the step-by-step fashion in which reactions occur without resorting to overly teleological, anthropomorphic, or mechanistic metaphors—metaphors our students may confuse with the reality of what we seek to describe.

INSTRUCTIONAL ACTIVITIES: STUDENT

17-1. Solids Versus Solutions. In a clean, dry mortar, gently combine 0.2 g each of lead nitrate and sodium iodide. Observe. Now grind the mixture, at first slowly and then more rapidly. Dissolve 0.2 g samples of the two substances in 10 ml water. Combine the two solutions. Observe. Begin thinking about accounting for the similarities and differences between this reaction occurring dry versus in solution. Write a correctly balanced equation for the overall reaction. Repeat this assignment but substitute structural formulas for the usual molecular formulas. Identify each inter- and intra-molecular bond that breaks or forms during the reaction.

Why is the reaction faster in solution? Why doesn't the combined solution need to be mixed, shaken, or stirred in order for the reaction to occur in contrast to the effect of the grinding of the solid system?

Repeat this entire exercise but substitute potassium iodide for the sodium iodide. If you see differences in the way the reaction occurs or fails to occur, to which specific bonds is it most logical to turn to find an explanation of these differences? What information would you seek were you to predict how the reaction would go were you to replace the original sodium iodide with lithium iodide? Cesium iodide?

17-2. Guessing or Hypothesizing the Mechanism of a Reaction. On the check-out table, you will find a saturated solution of cobalt(II) chloride hexahydrate. Transfer 6 to 8 ml of this solution to a small test tube. Add as large a lump of anhydrous calcium chloride as will go through the mouth of the test tube. Observe paying special attention to color changes within the system.

With or without joining forces with your lab station neighbors, think up as satisfactory explanation as you can of what you have observed. Consult handbooks or other references to back up your best hypothesis. Plan and carry out experiments to prove or disprove your hypothesis. Hint: Dip a strip of filter paper in the saturated cobalt chloride solution and then heat to dryness in an evaporating dish.

INSTRUCTIONAL ACTIVITIES: TEACHER

17-T1. The Traditional Iodine Starch Clock Reaction. Although we still hear teachers say, "I can never get that demonstration to work," few reactions seem to be more effective in confronting our students with the observations needed to bring out the desired mental exercise. In effect, the total system introduces several "competing" reactions into a common environment. In reverse order, the abruptly obvious reaction involves iodine molecules interacting with starch—a reaction that cannot occur until iodide ions are oxidized to iodine molecules. The oxidation of iodide ions to molecular iodine, however, is blocked by the concurrent interaction between iodate and hydrogen sulfite ions in an acidified environment. It is this "rate determining step" that lends itself so nicely to being made to fluctuate through changes in temperature and concentration.

Directions for preparing the solutions which reasonably careful lab assistants can follow are: Prepare solution A by dissolving 2.0 g potassium *iodate* in 1000 ml distilled water. Bottle, label, and store. Prepare solution B by placing 500 ml distilled water in a 1000 ml beaker or flask and bring to boil. Stirring constantly, dissolve 2 g of *soluble* starch. Then add 0.4 g sodium *hydrogen sulfite* and 5 ml of 1 M sulfuric acid. Add enough water to bring the total volume to 1000 ml. Bottle, label, and store.

These solutions should not be prepared more than a day or so before their scheduled use. Usually, if the reaction time is too slow it can be speeded up by adding a drop or two of the sulfuric acid. So that your students may gather a full array of data within a single class period, plan ahead to have 5 100 ml graduates filled with the iodate solution in a range of concentrations, for example, 100, 80, 60, 40, and 20 per cent of the original concentration. Similarly, arrange 5 100 ml graduates filled with the iodate solution and placed in water baths held at a range of temperatures, say, from that of an ice bath to 50°C or so.

These arrangements permit combining equal volumes of the iodate and starch solutions under reasonably standardized conditions whereby the time required for the reaction to occur can be determined. Also, sufficient data can be gathered during the class period and still leave time for data analysis, including preparing a graphed exhibit of the data. This analysis will involve these equations:

$$IO_3^- + 3HSO_3^- \longrightarrow I^- + 3SO_4^{-2} + 3H^+$$
$$5I^- + 6H^+ + IO_3^- \longrightarrow 3I_2 + 3H_2O$$
$$I_2 + HSO_3^- + H_2O \longrightarrow 2I^- + SO_4^{-2} + 3H^+$$

If approached through a bond-by-bond breaking and forming analysis, the function of the rate determining step can be emphasized—an approach that is supported dramatically by the abruptness of the color change which marks the final step of the reaction.

Hopefully, this instructional activity will clarify the roles of concentration and temperature in affecting the rate of a reaction. If we include clarification of the role of the sulfuric acid in the above activity, our students should be ready for confrontation with puzzling situations in which the "breakthrough" will be achieved by recognizing the effect of some factor adequate to affect the reaction rate.

17-T2. Factors Affecting Reaction Rates. Select 5 bottles or flasks of approximately 100 ml capacity and with mouths permitting stretching rubber balloons over the opening, 100 ml volumetric flasks do nicely (Figure 17-1). Place 50 ml of 6 m HCl in the first and equal volumes of 3 M, 1.5 M, 0.75 M, and 0.38 M hydrochloric acid in each of the other flasks. In each of 5 balloons place 2 g of mossy zinc. Without spilling any of the zinc into the flasks, stretch the mouths of the balloons over the mouths of the flasks. At an announced time signal, shake the zinc into the 5 concentrations of acid and record the relative promptness with which the balloons are sufficiently inflated to assume erect positions.

Parallel this demonstration with a second set of 5 flasks each containing 50 ml of 1.5 M hydrochloric acid but this time add decreasing volumes of approximately saturated copper sulfate solution to the acid before adding the zinc. Add 10 ml of the copper solution to the first, 5 ml to the second, and continue reducing the volume by one-half. Try to lead the discussion of the results from this activity to question the effects from diluting the acid by adding the copper solution.

Assuming that the class includes that type of student who delights in identifying inadequate controls in teacher demonstrations or otherwise spotting instances of questionable procedure in laboratory exercises, plan to reward the "troublesome" student by saying,

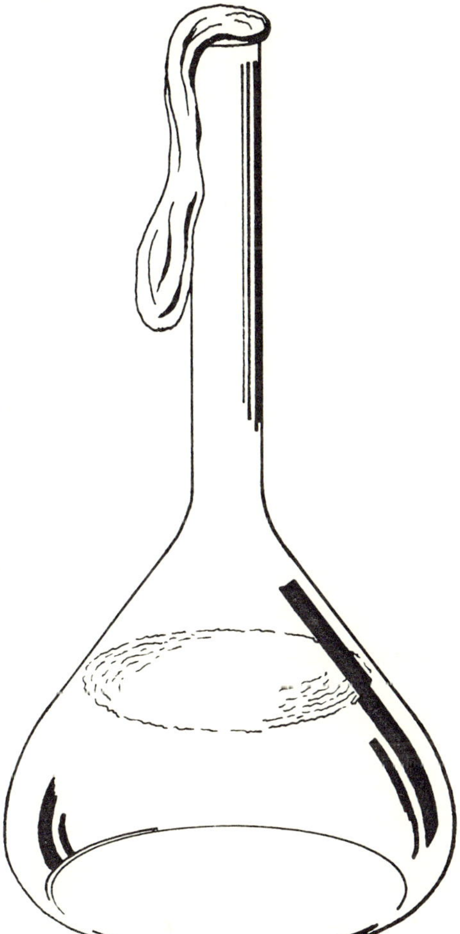

Figure 17-1

"Well, I happen to have a third set of flasks and balloons on hand and some solid $CuSO_4 \cdot 5H_2O$." The intent here is to confront the students with observations which suggest a catalytic rather than stoichiometric role for the copper ions in the reaction.

17-T3. Autocatalysis. Fill a large test tube one-third full with 3% hydrogen peroxide containing a few drops of 18 M sulfuric acid. Alerting the students to time the color change, assuming a color change will occur, add a drop at a time 0.01 M potassium permanganate. Through discussion, bring out that the concentration of the acidified hydrogen peroxide must certainly be diminishing as the reaction continues. Question the apparent contradiction of the usual relationship between concentration and reaction rate. Challenge your ability to nurture the "creation" of hypotheses in the minds of your students by seeing if they can come up with the hypothesis that a product of the reaction, in this case Mn^{+2} ions, might turn out to be a catalyst for the initial reaction.

17-T4. A Slightly "Sneaky" Mechanism. Arrange a series of 5 or more 250 ml beakers each containing 1 M sulfuric acid. In one beaker, place a zinc strip; a copper strip in a second; zinc and copper strips but not touching each other in a third; let them touch in a fourth; and a zinc strip with a loosely coiled length of platinum wire around it in the fifth. Station a student close to the beakers and have him give a running account of his observations throughout a 10 to 15 minute interval.

Announce that you have provided the additional beakers containing more of the acid and that you have a supply of other metals (some above and some below copper on the E° scale) and that you will be willing to conduct additional "experiments" upon request—experiments intended to help clarify the mechanism of the reaction between zinc and the sulfuric acid. Aim for such an idea as, "If the electrons being given up by the zinc can do so, they will go elsewhere than the zinc strip to send hydrogen molecules bubbling from the system and thus avoid protecting the zinc from the action of the acid with a blanket of hydrogen bubbles."

Audio-Visual Aids

17-AV1. "Catalysis." 10 min. Britannica. Nature and theory of.
17-AV2. "Catalysis." Color. 17 min. Chems. Mechanism of as illustrated

by sulfuric acid catalyzing decomposition of formic acid and the effects of peroxidase.

17-AV3. "Catalysts at Work." Color. 41 frames. FOM. Action of.

17-AV4. "Chemical Bonding." 20 min. Chems. Applies the concepts developed in the film, "The Hydrogen Atom," to the formation of a molecule of hydrogen.

17-AV5. "Chemical Kinetics." 44 frames. FOM. Uses several reactions to illustrate collision theory, activated complex, and the general concept of reaction mechanism.

17-AV6. "An Introduction to Reaction Kinetics." Color. 13 min. Chems. Activation energy, collision geometry, equilibrium.

17-AV7. "Mechanism of an Organic Reaction." Color. 20 min. Chems. Uses the hydrolysis of an organic ester to show structure of reactants, products, and intermediate particles.

17-AV8. "Molecular Motions." Color. 20 min. Chems. Illustrates translational, rotational, and vibrational motions during phase and molecular changes accompanying absorption and radiation of energy.

17-AV9. "Rate of Reaction." Color. 24 min. Coronet. Factors influencing reactions with markedly different reaction rates.

17-AV10. "Velocity of Chemical Reactions." 11 min. Chems. Factors influencing.

17-AV11. "Vibrations of Molecules." 11 min. Chems. Vibration types in simple molecules.

17-AV12. "Why Are Atoms Unpredictable?" 30 min. Indiana University Statistical probability of atom behavior.

Ideas for Additional Instructional Activities

17-AD1. Campbell, J. Arthur, *Why Do Chemical Reactions Occur?* Englewood Cliffs, New Jersey: Prentice-Hall, 1965. Involves the roles of energy and randomness in discussions of why some changes occur and others do not.

17-AD2. Eyring, Henry, and Edward M. Eyring, *Modern Chemical Kinetics.* New York: Reinhold, 1963.

17-AD3. Dence, Joseph B., Harry B. Gray, and George S. Hammond, *Chemical Dynamics.* New York: W. A. Benjamin, 1965. Introduction to the systematic study of reactions and reactivity.

17-AD4. Nicholson, Douglas G., and Thomas H. Edwards, "A Demonstration Model for Transport Reactions," *Chemistry,* 39: 30–31. October, 1966.

17-AD5. Taube, Henry, "How Do Redox Reactions Take Place," *Chemistry,* 38: 18–23. March, 1965. Discusses several mechanisms that have been proposed for oxidation-reduction reactions in aqueous solutions.

17-AD6. Videen, Tom, "Burning Rates of Candles," *Chemistry,* 39: 26–27. August, 1966. An exercise involving combustion rates associated with changes in candle and/or wick diameters.

17-AD7. Zwolenik, James J., "Flash Photolysis: Catching Intermediates on the Fly," *Chemistry,* 40: 17–20. Brief discussion of photochemistry and the use of flash photolysis in reaction mechanism studies.

18

EQUILIBRIA SYSTEMS

Many of the reactions which students encounter in their introduction to chemistry go to completion, that is, the reactants are converted completely to products. Seldom are students confronted with the idea that a chemical system may exist in which there is a kind of standoff between the reactants and products—systems in which significant concentrations of reactants and products coexist. As the course unfolds, however, adequate appreciation of the dynamics of chemical systems at equilibrium becomes increasingly important.

In many respects, man approaches the chemical systems he hopes to manage or exploit when they are in a state of equilibrium albeit a state in which the concentrations of the products may exceed greatly the concentrations of the reactants. This static state must be rendered dynamic by one means or another before new rearrangements of the atom components can be brought off. And it is the intent of the chemist to direct the reaction in such a way as to maximize the yields of desired products in the course of the reaction returning to a new state of equilibrium.

The chemistry of living systems is closely linked to delicate balances between the concentrations of substances in closed chemical systems. Knowledge and appreciation of the fantastically complex yet precisely controlled physico-chemical processes which add up to the life process requires us to deal with equilibria systems. It is inescapable that the input or outgo of energy is one of the stresses most likely to upset the equilibria of biochemical systems. We see this in the continuously varying input of solar energy into each terrestrial habitat. We also see this in the life activities of each individual. Hopefully, knowledge of the factors which seem to be adequate to affect biochemical systems at equilibria will produce the behavior whereby each student enjoys the physical and mental health needed to live life ever more abundantly. And although these ideas stem more directly from the reactions encountered in later chapters, particularly biochemistry, the groundwork must be laid in an enlightened comprehension of the dynamic nature of chemical systems.

INSTRUCTIONAL ACTIVITIES: STUDENT

18-1. Comparing Sets of Chemical Systems. Using small test tubes, combine 4 ml each of 1 M lead nitrate and sodium iodide. While waiting for the precipitate to settle (centrifuge if convenient), write an equation for this reaction. Anticipate what should happen if (1) a drop of additional lead nitrate is added to the clear liquid and (2) if a drop of additional sodium iodide is added. Verify if you are in doubt.

Using small test tubes, combine 4 ml each of 0.01 M ferric nitrate and 0.01 potassium thiocyanate. Assuming the deep red product to be $FeCNS^{+2}$, write the equation for this reaction. Set this test tube aside to use as a reference and prepare a similar combination of the two solutions. Divide this solution into two test tubes. To one add 2 drops of the ferric nitrate and to the other 2 drops of the thiocyanate solution. Identify any differences here with your observations when additional drops of the solutions were added to the previous lead nitrate-sodium iodide system.

Verify the accuracy of the calculations shown on the chalkboard and then add the required volumes of stock bottle hydrochloric and acetic acid needed to produce 100 ml of 0.01 M solutions of the two acids. Determine the pH of the two solutions. Determine the number of drops of 6 M NaOH needed to neutralize the two acid solutions to phenolphthalein end point. Write equations for the ionization of each of the two acids as well as for the reactions involving their neutralization with the sodium hydroxide solution.

Keeping in mind all of the observations involving these three sets of systems, how many of the apparently puzzling aspects would be explained if we can assume that some kinds of reactions do not go to completion, that is, portions of the initial reactants can coexist with the products? On the basis of this assumption, what would happen to the density of the color in a $FeCNS^{+2}$ system if you added one or more of the following substances: $FeCl_3$, NH_4CNS, NH_4Cl, KCl?

18-2. Putting Equilibria Systems Under Stress. Put a large ice cube or two into a foam plastic cup or calorimeter cup. Fill the cup three-fourths full with water. Stir as you record the temperature until a constant temperature suggests there is an equilibrium between water, liquid, and water, solid, in the system. Now add another ice cube and then record the temperature for a few minutes. Now add some water that has been warmed twenty or so degrees above room temperature and then record the temperature for several minutes.

If we assume that a constant temperature can be evidence of equilibrium, what happens to an ice-water equilibrium system when we bring it under stress by increasing the quantity of either of the two components?

Prepare a saturated solution of sodium chloride by adding approximately 50 g of salt to 100 ml water. Stir thoroughly until you believe you have established an equilibrium between salt, dissolved, and salt, crystalline. Using a pipette, transfer 10.0 ml of the saturated solution to an evaporating dish and evaporate over a boiling water bath. While this sample is evaporating, add approximately 20 more grams of salt to the original saturated solution. Again, stir thoroughly and then transfer 10.0 ml of the saturated solution to an evaporating dish. Proceed to evaporate both samples to the same degree of dryness and determine the weight of salt that must have been dissolved in each of the two samples. Keeping the confidence limits of your balance in mind, and the precision of your pipette, did adding additional salt to the original solution equilibrium affect the concentration of salt in the saturated phase of the system?

To 100 ml distilled water in a 150 ml beaker add 3 drops of 17.4 M acetic acid. Stir until you can assume an equilibrium has been established between hydrogen and acetate ions in the system. Use a pH meter or pH papers to estimate as precisely as you can the concentration of hydrogen ion in the system. Now add 3 more drops of the 17.4 M acid. What is the effect from apparently doubling the quantity of acid added to the water? Continue this investigation by adding two more 3 drop increments.

Contrast the responses of the three systems; ice-water, salt, dissolved-salt, crystal, and hydrogen ion-acetate ion, to the kinds of stresses you have applied. Which of the three systems seems to be the most "responsive" to stress? What do the two less "responsive" systems have in common that is not present in the most "responsive" system?

18-3. The Common Ion Effect. Put 100 ml distilled water in each of 4 150 ml beakers. To the first, add 4 drops of 14.8 M ammonium hydroxide; 8 drops to the second, 12 to the third, and 16 to the fourth. Determine the pH of each system as precisely as your facilities permit. Now, approximately double the quantity of ammonium ion present in each system by adding 0.6 g of solid ammonium chloride to the first, 1.2 g to the second, 1.8 g to the third, and 2.4 g to the fourth. What are the effects on the pH of each system?

To help you answer this question or, rather, to help you account for your observations, be sure to study the equations for each system both before and after you have brought them under stress by increasing the concentration of ammonium ion.

For further investigation of the common ion effect, half fill 3 test tubes with saturated solution of silver acetate. To one test tube, add a few small crystals of solid silver nitrate. Add a few crystals of sodium acetate to the second test tube. Use the third test tube as a "control" lest unforeseen changes in the laboratory environment may occur.

INSTRUCTIONAL ACTIVITIES: TEACHER

18-T1. Stresses Applied to Gaseous Systems at Equilibrium. Fit a reasonably thick walled 500 ml or larger flask with a 3-hole rubber stopper. Place a thermometer in the stopper so the bulb extends well toward the center of the flask. Using a second hole in the stopper, arrange to have water enter the flask from the water mains. Use the final hole in the stopper to connect a closed tube manometer to the system.

Open the tube leading to the manometer and allow a little water to enter the flask. Attach the manometer and encourage your students to identify the relative interaction of the three factors, pressure, volume, and temperature, acting on the trapped parcel of gas in the flask. When all these factors seem to be at equilibrium, allow some water to run into the flask. Discuss as precisely as your set-up permits the apparent response of each of the other two factors when the system is brought under stress by diminishing the volume of the air in the flask. Assuming you do not exceed the pressure limits of the set-up, introduce additional quantities of water and repeat the discussion.

Audio-Visual Aids

18-AV1. "Equilibrium." 24 min. Color. Chems. The dynamic nature of and determination of equilibria constants.

18-AV2. "The Equilibrium Constant." 42 frames. FOM. Investigation of physical and chemical equilibria as dynamic phenomena.

18-AV3. "Ionic Equilibrium." 16 min. Coronet. Ionic equilibrium as illustrated by cupric bromide, common ion effect as illustrated by acetic acid and the acetate ion, etc.

18-AV4. "Ionization and Ionic Equilibrium." Color. 15 min. Indiana University. Pictorial and quantitative presentation of dissociation, common ion effects, buffering, and similar phenomena.

18-AV5. "Le Chatelier's Principle." 44 frames. FOM. Operation of the principle under varying sets of conditions.

Ideas for Additional Instructional Activities

18-AD1. Allen, J. A., *Energy Changes in Chemistry*. Boston: Allyn and Bacon, 1966. Using the concepts of equilibria and free energy to predict chemical reactions.

18-AD2. Banks, James E., *Chemical Equilibrium and Solutions: A Programmed Introduction*. New York: McGraw-Hill,

18-AD3. Barrow, Gordon M., *et al, Chemical Equilibria*. New York: W. A. Benjamin, 1967. Volume IV of the series, "Understanding Chemistry," consisting of approximately 200 programed instruction frames. Other titles in the series are I, Chemical Quantities; II, Chemical Bonding; III, Chemical Reactions; and V, Chemical Systems.

18-AD4. Bauman, Robert P., *An Introduction to Equilibrium Thermodynamics*. Englewood Cliffs, New Jersey: Prentice-Hall, 1966. A framework to which the foundations and implications of thermodynamics can be attached.

18-AD5. Carnell, P. H., and R. N. Reusch, *Molecular Equilibrium,* Philadelphia: W. B. Saunders, 1963. A programed course consisting of 564 question-answer frames.

18-AD6. Clak, John N., "Dynamic Equilibrium an Alcohol-Iodine System," *Chemistry,* 40: 50. July-August, 1967.

18-AD7. Ellis, George, *et al,* "Determining the K_A of Weak Acids," *Chemistry,* 38: 28–29. August, 1964. Uses a pH meter to determine the K_A of acetic acid.

18-AD8. Morris, Kelso B., *Principles of Chemical Equilibrium.* New York: Reinhold, 1965. For the general reader of college chemistry.

18-AD9. Schaff, John F., and Paul Westmeyer, "Dynamic Nature of Chemical Equilibrium," *Chemistry,* 41: 48–50. July-August, 1968.

18-AD10. Sienko, M. J., *Equilibrium, Freshman Chemistry Problems and How to Solve Them.* New York: W. A. Benjamin.

19

THERMOCHEMICAL PHENOMENA

Although energy transformations accompany nearly all chemical reactions, our students must rely on more or less indirect observations in order to describe or understand the energy aspects of reactions. Heat radiated from or absorbed into chemical systems provides just about the most direct evidence that a chemical change is accompanied by energy transformation. But nearly all of our students must learn how to interpret this evidence. They are usually able to observe temperature differences between the initial and final states of chemical systems but they have no understanding of how temperature changes are tied to changes in heat content.

Once our students have learned how closely energy transformations are related to chemical events they invariably raise "how" and "why" questions. These questions not only take us quickly to the very nub of chemistry but they also nudge us toward the interface between chemistry and philosophy. Some teachers enjoy the challenge of trying to fend off students who are unsatisfied with such statements as, "chemical systems tend toward minimum energy and maximum randomness." Others don't. But our students certainly have the right to expect us to make sense out of such terms as "entropy" or "change in free energy." The problem seems to be to emphasize the significance of such concepts and to open up glimpses of how they lead toward ever increasingly precise yet simple descriptions of reactions—glimpses adequate to keep curiosity alive until our students acquire the related knowledge that enables them to use the language that must be used to avoid confusing facts with fiction, chemistry with philosophy.

INSTRUCTIONAL ACTIVITIES: STUDENT

19-1. Temperature Versus Heat Content. Place 100 g of water that has been cooled to 15°C. in a styrofoam cup. Add 100 g of water that has been heated to 75°C. Record the temperature of the mixture and calculate the number of calories of heat gained by the cold portion and lost by the hot portion.

Repeat this exercise but this time use 50 g of the cold and 150 g of the hot water. Or 50 g of the hot and 150 g of the cold.

Use the results from all of your work to reveal the imprecision in such a statement as: There surely isn't much heat left in this cup of coffee.

Punch several small holes in the bottom and sides of a styrofoam cup. Put 100 g of aluminum shot, punchings, or tacks in the cup and stand the cup in 15°C. water until the aluminum has cooled to approximately this temperature. Lift the cup and allow the water to run off then add the aluminum to 100 g of 75°C. water. Calculate the number of calories gained by the "cold" aluminum and lost by the "hot" water. Continue this investigation by changing the amounts and temperatures of the aluminum versus the water.

Use the results from this exercise to discuss the imprecision in such a statement as: Bodies of land lose their heat more rapidly than does the surrounding water.

19-2. Temperature Changes with a Difference. Using styrofoam cup calorimeters, add 5 g of drain cleaning compound (or sodium hydroxide) to 50 ml of room temperature water. Calculate the number of calories gained or lost by both the solid and the water. If you are uneasy about your data, live with them for the time being. Now add 5 g of ammonium nitrate to 50 ml of room temperature water. Again, calculate the number of calories gained or lost by the solid and the water. Bring both of the solutions to approximately the same temperature and then combine them. Again, calculate the calories gained or lost by each of the two original solutions.

How do your data from this instructional activity differ from or seem to contradict the orderly relationships which seemed to prevail in activity 19-1? Suppose someone says that in 19-2, chemical changes took place but this was not the case in 19-1. What would be the effect on the validity of this hypothesis were you to evaporate a sample of the drain cleaning or ammonium nitrate solutions and find crystals of the original compounds? Similarly, what would be the effect on this hypothesis if, upon evaporating a sample of the combined solutions, you find a substance or substances other than the original compounds?

19-3. Properties of Metals Before and After Being Heated. The goal of this activity is to compare the properties of metals before and after being heated and the relative effects of rapid versus slow cooling. Obtain short lengths of several different kinds of wire. Note how easily each kind of wire bends. Heat each kind of wire to redness and note, after cooling, whether it makes a difference if the metal is cooled rapidly by plunging into cold water, cooled slowly by allowing it to cool in air, or even more slowly by plunging the hot metal into hot oil.

Use your observations to judge the validity of this hypothesis: If metals are cooled slowly, the individual particles of the metal tend to arrange themselves in definite crystal lattices and the cooled metal will have different properties than it has when the particles are frozen into no particular arrangement.

19-4. Heat to Start Versus Heat to Continue a Reaction. Fill the rounded portion of the bottom of a test tube with ammonium dichromate. Use a burette clamp to secure the test tube to a ring stand. Be sure to tilt the test tube and avoid pointing it toward yourself or a classmate. Heat the bottom of the test tube but remove the flame just as soon as you see evidence of a chemical change occurring. If no changes continue after removing the flame, resume heating but remove the flame immediately when you see any reaction resume. Keep doing this until you are fairly sure that no further changes can be accomplished by further heating.

When interpreting your observations, ask yourself such questions as: What changes did I see take place in the substance in the test tube? How many of the changes seemed to be caused by the heat coming from the burner? Did the burner provide all of the heat that seemed to accompany the total reaction? If not, how can the unaccounted for heat be accounted for? What would happen if you closed the test tube with a one-hole rubber stopper fitted with a glass tube that was bent in such a way that any gases produced by the reaction would be bubbled through water?

19-5. The Heat Additive or Hess's Law. Using a styrofoam cup or a regular calorimeter, calculate the calories of heat accompanying the interaction of 2.0 g of solid sodium hydroxide being dissolved in 100 ml of water. To do this, add the sodium hydroxide and record the temperature at 30 sec. intervals until the maximum temperature rise occurs. Assume that the heat capacity of the resulting solution does not differ significantly from that of water. On this basis, the calories of heat lost by the reactants will equal the temperature change multiplied by 100. Repeat this procedure but use 0.5 M HCl in place of the water.

The reaction you have just brought off may be shown by the equation:

$$NaOH_{(s)} + H^{+}{}_{(aq)} + Cl^{-}{}_{(aq)} \longrightarrow H_2O + Na^{+}{}_{(aq)}\ Cl^{-}{}_{(aq)} + x\text{cal}$$

Ask yourself where the heat might have come from that raised the temperature of the water. The chloride ions are unlikely suspects; they could well have stayed clear of the whole reaction. All other particles, however, are suspect. The sodium hydroxide crystal structures had to break down to release sodium and, probably, hydroxide ions. This change may wrap up the total role of the sodium ions but the newly released OH ions seems to have joined H ions to form water molecules.

This kind of reasoning leads us to such an hypothesis as: The heat of this reaction came from breaking down the sodium hydroxide crystal structure and/or the union of H and OH ions to form water molecules. But don't ignore another possibility. One of these reactions may have released and the other absorbed heat. All our thermometer tells us is that the net reaction was exothermic. Fortunately, we can pin down the temperature change by carrying out each of the two suspect reactions separately.

To do this, put 100 ml of water in a styrofoam cup and calculate the calories of heat gained or lost when 2.0 g of solid sodium hydroxide are dissolved in the water. Arrange to have the temperature of this solution to be the same as that of 100 ml of 0.5 M hydrochloric acid. Now combine the acid and the base solutions and calculate the calories of heat lost or gained during this reaction.

When a reaction occurs in steps, the heat gained or lost by the total reaction is the algebraic sum of the gains and losses of the individual steps. Do your data agree with this generally accepted state of affairs? Keep in mind that the precision of your instruments, especially the thermometer, must be considered when answering this question.

For another reaction whereby you can observe the action of the heat additive relationship, dilute 40 ml of concentrated $NH_{3(aq)}$ with 100 ml water. Record the temperature of this solution and immediately pour over 16 g of anhydrous copper sulfate in a thermal cup. Record the maximum temperature change and calculate calories. To carry out the same overall reaction a step at a time, first dissolve the copper sulfate in the water, then add the ammonia, and record the temperature changes separately for the two steps. Be sure to write the equations for each of the reactions so you can interpret your data effectively.

INSTRUCTIONAL ACTIVITIES: TEACHER

19-T1. Interactions of Electrode Potential Difference, Delta-H, Delta-G, Delta-S, and Entropy. One big advantage of the demonstration format permits time to be budgeted so as to insure that our students receive a complex set of ideas in such a way that their interrelatedness can be emphasized. Assuming adequate advance preparation, all of the following bits and pieces can be brought off within one class session or laboratory period.

Add approximately 1 g of zinc dust to 200 ml of 0.20 M silver nitrate in a thermal cup. Record maximum temperature rise and, thence, delta-H for this reaction. Scale up to the traditional kilocalories per mole.

To observe the same reaction from the viewpoint of electrochemical changes, prepare an electrochemical cell with 60 ml of 1.0 M silver nitrate in the porous cup and 100 ml of 1.0 zinc nitrate in the beaker. Add the electrodes (silver and zinc) and note electrode potential difference. Note that, in effect, this electrochemical cell permits zinc atoms to go into solution only if silver ions come from solution as silver atoms, that is $Zn_{(s)} + 2Ag^{+}{}_{(aq)} \longrightarrow Zn^{++}{}_{(aq)}, + 2Ag_{(s)}$.

Delta-H for the reaction when determined as in the first step of this demonstration should be close to 86.8 kcal/mole. Potential difference for the zinc-silver cell should be close to 1.56 volts. To calculate the energy involved were one mole of zinc atoms to "move" two moles of electrons over to as many silver ions, change the faradays to the equivalent 193,000 coulombs and, hence, to volt-coulombs. Since one volt-coulomb equals one joule, and one calorie equals 4.18 joules, it can be shown that when one mole of zinc atoms gives up two moles of electrons to two moles of silver ions, 72.1 kcal would be lost by the system, a measure we call delta-G.

Relying primarily on the thermometer to tell us about the energy changes in this reaction, we found that the system lost 86.6 kcal per mole of reactants. Relying primarily on the voltmeter, we can account for an energy loss of only 72.1 kcal. What about the "missing" 14.5 kcal, assuming the usual degree of faith in the law of conservation of energy? The answer lies in what our instruments actually measure. Delta-H is a measure of the total or net heat loss when the reaction occurs. Delta-G is the fraction of this heat loss that was converted to electrical energy and, in effect, is available to do work outside the system in which the reaction occurred.

The "missing" energy is described as the change in entropy of the system or, delta-S. From some points of view, entropy changes are associated with the relative order or disorder (randomness) of the system. The inference, somewhat philosophical, is that all systems move relentlessly toward states of increasing randomness with concurrent loss of energy. It is as though the energy make-up of any system cannot be jiggled in any way without allowing the system to slip, at least for a time, down the entropy gradient. Actually, to describe with truly satisfying precision the *apparent* disappearance of our "missing" 14.5 kcal can be something to look forward to after we have completed our introductory courses in chemistry and move on to advanced study of the big ideas involved in delta-G, delta-S, and their relationship to delta-H.

Our study of this one chemical system in which zinc atoms and silver ions interacted calls attention to several of the kinds of things we associate with whether or not a reaction "goes"—whether or not a set of potential reactants do react or simply coexist with nothing happening. We have seen that a reaction is favored (1) whenever the heat content of the reactants exceeds the heat content of the products which might result from a reaction, (2) whenever electrons can transfer from higher to lower energy state, and/or (3) whenever the entropy of the system can undergo increase, that is, whenever the system, by reacting, can become less ordered. When a potential reaction fails to come off, it is as though there is a kind of standoff between one or another of these factors opposing the others.

The replacement of copper ions by zinc may be substituted for the zinc-silver ion interaction in this instructional activity. Delta-H for this reaction has been reported to be 49.05 kcal.

Audio-Visual Aids

19-AV1. "Energy and Reaction." Color. 14 min. McGraw-Hill. Role of energy in making and breaking chemical bonds.

19-AV2. "Mechanism of an Organic Reaction." Color. 20 min. Modern Learning. From initial to final states by way of activated complexes.

19-AV3. "Thermochemistry." 44 frames. FOM. Representative calculations to illustrate Hess's Law of Heat Summation.

Ideas for Additional Instructional Activities

19-AD1. Alyea, Hubert N., "TOPS in Chemistry," *The Science Teacher,* 29: 14–21. October, 1962. Methods and devices for projecting demonstrations involving thermochemical phenomena.

19-AD2. Armstrong, George T., "The Calorimeter and Its Influence on the Development of Chemistry," *Journal of Chemical Education,* 41: 297–306. June, 1969.

19-AD3. Austin, Leonard G., "Fuel Cells," *Scientific American,* 201: 72–78. October, 1959. Conversion of chemical energy directly to electricity.

19-AD4. Davidson, D. L., "Laboratory Uses of Heat-Insulating Blocks," *Journal of Chemical Education,* 22: 38. 1945. Fitting beakers into blocks of heat-insulating material produces quite efficient calorimeters.

19-AD5. Haggin, Joseph, "Thermocouples for Measuring Temperature," *Chemistry,* 39: 34–35. December, 1966. Making and calibrating copper-constant in thermocouples.

19-AD6. Hill, Derek L., Stephen J. Moss, and Robert L. Strong, "Heat of Reaction in Aqueous Solution by Potentiometry and Calorimetry," *Journal of Chemical Education,* 42: 541–544. October, 1965. Finding delta-H from potentiometry and checking with calorimetry using a metal displacement reaction.

19-AD7. Livingstone, R., and J. J. Lingane, "An Experiment Illustrating the Relation Between Electromotive Force and the Equilibrium Constant," *Journal of Chemical Education,* 15: 320. 1938. Showing the relationship between electromotive force and the equilibrium constant (delta-G) using the hydroquinone-silver nitrate system. Typical data are included for delta-G°, delta-H°, and delta-S.

19-AD8. Mahan, Bruce H., "A Simple Ice Calorimeter," *Journal of Chemical Education,* 37: 634. 1960. And its use in determining enthalpies of reaction between magnesium and hydrochloric acid, formation of magnesium ion, and solution of the magnesium ion so as to bring out addition of enthalpies.

19-AD9. Owen, Brewster, "Latent Heat of Evaporation," *The Science Teacher,* 21: 153. April, 1954.

19-AD10. Pattison, D. B., J. G. Miller, and W. W. Lucasse, "Simplified Calorimetric Studies of Various Types," *Journal of Chemical Education,* 20: 319. 1943. Significance of delta-H and data obtained from thermochemical investigations.

19-AD11. Porter, George, "The Laws of Disorder," *Chemistry,* 41: 23–25, May, 1958. Describes a television series of lectures with demonstrations.

19-AD12. Siegel, Bernard, "Relationship of Exothermicities of Compounds to Chemical Bonding," *Journal of Chemical Education,* 40: 308. June, 1963.

19-AD13. Stong, C. L., "On Equipment to Study Freezing and Growing Crystals of Salt," 128–131. January, 1968.

19-AD14. Wilhoit, Randolph C., "Recent Developments in Calorimetry," Part I, "Introductory Survey of Calorimetry," *Journal of Chemical Education,* 44: A571; Part II, "Some Associated Measurements," 44: A629, A685; and Part III, "Some Specific Types of Calorimeters," 44: A583. 1967. Thermometry, thermocouples, thermistors, quartz crystal thermometers.

19-AD15. Wilson, Mitchell, "Count Rumford," *Scientific American,* 203: 158–168. October, 1960. The death of caloric and the birth of molecular motion.

19-AD16. Ziman, John, "The Thermal Properties of Materials," *Scientific American,* 217: 180–188. September, 1967. The phenon concept, a particle-like packet of waves that can travel through a solid.

20

CARBON CHEMISTRY

Organic chemistry confronts today's teachers with a special challenge. Student interest in this branch of chemistry is often whetted by widespread, often sensationalized new accounts of new drugs and medicines, plastics, herbicides and insecticides, food additives, textile dyes or conditioners, and other products of modern industrial organic chemistry. In one way, this is good. It also creates pedagogical problems. Students who are set for the exciting chemistry of exotic compounds may feel let down when their introduction to the chemistry of carbon and its compounds appears to deal with totally prosaic compounds. At the same time, they may look upon even these simple substances as having singularly complex formulas or highly involved properties.

The problem is to exploit student interest in the excitingly practical aspects of organic chemistry and to use this interest to help them learn the fundamental principles and concepts of the topic. Students should see the values in logical systems for classifying and naming the almost countless compounds of carbon. Equally important are the general concepts involved in the synthesis of compounds, precise measurements of their properties, and the identification of the relationships between these properties and whatever functional groups exist in their molecules.

The arrangement of atoms within molecules, rather than the kinds and numbers of the atoms forming the molecules, receives major attention when studying the chemistry of carbon's compounds. For this reason, organic chemistry provides a good opportunity to review and extend our student's ideas and concepts regarding atomic structure and molecular architecture.

INSTRUCTIONAL ACTIVITIES: STUDENT

20-1. Structural Formulas in Styrofoam. Although you may feel a vague uneasiness when you use physical models to simulate situations which are fundamentally abstract, one or another kind of "ball and stick" system of models can be quite useful in visualizing the structure of organic compounds. And it is the structure of their molecules that sheds the most light on the reasons certain substances have the properties they do. In fact, only a few kinds of atoms can be joined together in many, many different ways and, thus, produce many, many different kinds of substances.

Consult the chart that shows the relative sizes of the atoms. Select from the supply of foam plastic balls, 10 carbon, 10 oxygen, and 25 hydrogen "atoms." Also take a supply of the pipe-cleaner "valence bonds." If necessary, you can color additional balls by shaking them in a large jar with a little dilute showcard pigment and scattering them to dry on newspapers. See the additional suggestions regarding the use of foam plastic balls for models of molecular structures in 5T-1.

Use your model atoms and valence bonds to visualize the relationships among the members of several homologous series of organic substances. For example, team up with the other students at your lab table and construct models of methane, ethane, propane, butane, pentane, and on into the higher number carbon atom hydrocarbons. Show how saturated and unsaturated hydrocarbons differ. Show how step-by-step oxidation of the hydrocarbons leads to the formation of other types of organic compounds, for example, alcohols, aldehydes, ketones, acids, or ethers.

Keep in mind that it is easy to allow this activity to become nothing more than an interesting little game. Don't forget that its purpose is to help you visualize the sometimes complex structure of organic molecules and, eventually, to help you understand the chemistry of these substances.

20-2. Organic Compounds from the Oxidation Series Point of View. Faced with the fantastically large number of different kinds of compounds containing carbon, the need is obvious for some scheme whereby compounds with similar properties can be grouped into, hopefully, a small number of families or classes of compounds. One such scheme is based on the degree of oxidation the molecules of certain groups of substances have undergone. A delightfully logical scheme begins with hydrocarbons and ends with carbon dioxide and water. Between these two extremes we find alcohols, aldehydes, acids, and, possibly, ketones or ethers and esters.

Obtain a supply of appropriately colored foam plastic balls and "valence bonds" ignoring for the time being that they look very much like short lengths of pipe-cleaners. Construct models first of methane, methanal, methanol, and methanoic acid. Keep these models on display and then construct models of propane, propanal, propanol, propanone, and propionic acid.

Recall the following rules involving oxidation numbers: (1) Assign oxygen the number -2. (2) Assign hydrogen the number $+1$. (3) The sum of the oxidation numbers equals the charge on the molecules (assumed to be zero). Apply these rules and determine the oxidation state of carbon in each of the molecules whose models you have built. Are there orderly relationships in these data?

Leaving the models but retaining the perspective they provide, place about 5 ml of methanol in a small test tube. Heat a length of copper wire coiled on one end until it glows red. Observe as it cools. Reheat until it glows red again and plunge it immediately into the vapors of methanol just above the liquid phase in the small test tube. Smell (*caution*) the vapors coming from the test tube and compare the new condition of the copper wire with its condition when it cooled in air. When interpreting your observations, here are some hints. Does the smell remind you of some of your biology laboratory exercises? If the copper in copper oxide is reduced, what must have happened to something else at the time?

What lesson could we teach you if we knew of other convenient to pull off reactions that would do to the formaldehydes of the previous reaction what the hot copper wire did to the methanol? Returning to your two series of models, what is the logical extension of this question?

20-3. Managing or Manipulating Organic Reactions. Although not totally different, organic reactions seem to call for different strategies than those which work successfully for the chemists who are interested in managing reactions involving primarily inorganic substances. Exaggerated for the sake of emphasis, inorganic reactions begin with a grand and glorious breaking up of the molecules of the reactants. And the more successfully all pre-existent bonds within molecules are fractured, the better the yield of the desired products.

But to manage organic reactions successfully, more elegant strategies seem to be called for. The goal becomes attacking a complex molecule only at the point where you want specific bonds to be broken or to open up so that a new atom or group of atoms will attach. And this strategy demands very precise knowledge regarding the relative strengths of all of the different kinds of bonds within the complex molecule. In metaphor, when an architect finds that a house doesn't quite fit his needs, he surveys the situation very carefully so as to bring about the desired remodeling with the least damage to the existent structure and at the lowest investment of time and money. This sort of strategy is illustrated in this instructional activity.

Build foam plastic models of acetone, CH_3COCH_3, and iodoform, CHI_3. Study these models and decide what would be the most efficient way to make iodoform from acetone. Now dissolve a pellet of NaOH in 2 ml of water and add several crystals of iodine. After the iodine is dissolved, add a few drops of acetone. Consult the *Handbook of Chemistry and Physics* for the properties of iodoform and decide whether or not you have some of this substance in the test tube.

For a second example of organic reaction management, build models of salicylic acid, 2-hydroxy-benzoic ($C_7H_6O_3$), and methyl alcohol. Play with these two models and decide how they could be used to make the ester, methyl salicylate. Now put the models aside and place 1 g of salicylic acid in a 100 ml beaker. Add 10 ml methanol. *Cautiously* and under the *hood,* pour about 5 ml of 18 M sulfuric acid down the side of the beaker. Heat gently under the *hood* for about 15 minutes. Cool the mixture and then *cautiously* add about 15 ml of water. Use your nose together with the *Handbook* for evidence that you have successfully synthesized methyl salicylate.

20-4. Deducing the Structure of Organic Molecules. At the check-out table you will find such hydrocarbons as hexane, cyclohexane, cyclohexene, benzene, toluene, etc. Build models of the molecules of each hydrocarbon. Now put about 10 drops of each hydrocarbon in small test tubes and then add about 20 drops of Br_2 in TTE to each test tube. Place a stopper in each test tube and shake thoroughly. What structural feature do you find in the molecules of each of the hydrocarbons in which the bromine color disappears?

Fit a 6-inch test tube with a one-hole stopper carrying a glass tube bent so that gas generated in the tube can be led to the bottom of a second but smaller test tube. Half fill the smaller test tube with Br_2 in TTE. Remove the stopper from the first test tube; drop in a small lump of calcium carbide, CaC_2; add about 5 ml of water; and quickly replace the stopper. Before looking up the formula for acetylene, see if you can use the information from your own observations to deduce the structural formula of the gas that bubbled through the small test tube. In general, what feature of organic molecule structure can be inferred if the substance decolors the bromine solution?

Pour about 15 ml 1.0 M silver nitrate into a large test tube and add 1.0 M ammonium hydroxide drop by drop until the precipitate that forms dissolves. Then add one drop of 1.0 M sodium hydroxide but if a precipitate reappears, add enough ammonium hydroxide to clear the solution. This reagent is called Tollen's solution and the active particle is $Ag(NH_3)_2^+$. CAUTION: This solution should not be stored because it is known to produce an explosive compound on standing.

Put about 5 ml of the Tollen's solution in a small test tube and add a few drops of formaldehyde. Mix quickly and let stand for several minutes. Assume that the observations you record are unique for all compounds containing the aldehyde group. Prepare 2% solutions of glucose and of sucrose. Use about 5 ml samples of these two solutions and equal volumes of the Tollen's reagent to test for the presence of the aldehyde group in the molecules of these two sugars. After mixing, set the test tubes in 95°C. water for about 5 minutes. Do either or both sugars contain aldehyde groups?

In general, what strategy is employed in developing an appropriate test for the presence of a particular functional group in the molecule of an organic substance?

INSTRUCTIONAL ACTIVITIES: TEACHER

20-T1. Cis-Trans Isomerism. As a concept, this form of isomerism may well be too sophisticated to be included in the introductory chemistry course. The two isomers, maleic acid and fumaric acid, illustrate the concept very conveniently, however, and in doing so, bring out additional ideas which tend to enhance appreciation of more fundamental concepts.

Show your students models of maleic acid (cis-butenedioic) and fumaric acid (trans-butenedioic), $HO_2CCH{:}CHCO_2H$. Demonstrate the relative solubilities of the two substances in water but, assuming your way of teaching resonates with this sort of approach, avoid identifying the two isomers. Confront your students with 3 beakers containing mineral oil at, say, 130°C., 160°C., and 190°C. Place small test tubes containing samples of the two isomers in each of these hot oil baths and emphasize the difference in melting temperatures.

Put about 6 g of maleic acid in a 100 ml beaker, dissolve in 10 ml of water, add 15 ml of 12 M hydrochloric acid, and cover with a watch glass. Using a 250 ml beaker as a water bath, heat the maleic acid-HCl system and boil until a solid appears in the 100 ml beaker. Cool under running water, filter off the solid using cold water to wash the crystals onto the filter paper, and dry under a heat lamp or on a thick asbestos mat. Show your students that the original material (maleic) differs from the final material (fumaric) in the same ways that the samples originally from different bottles differed.

Request your students to accept on faith that the HCl did not undergo any permanent "reshuffling" of its make-up during the apparent reaction in the small beaker. Entertain hypotheses adequate to account for all observations and inconsistent with none. If you can do so adroitly, use the models to help your students predict that one arrangement exceeds the other in creating hydrogen bonding. If you choose to deduce a mechanism for the isomeric shift, good luck.

Audio-Visual Aids

20-AV1. "Carbon and Its Compounds." Color. 11 min. Coronet. Introduces representative chain and ring compounds.

20-AV2. "The Chemistry of Carbon and Silicon." 44 frames. FOM. From viewpoint of ionization energy, electronegativity, macromolecular structures, chemical reactivity, etc.

20-AV3. "Functional Organic Groups." 40 frames. FOM. Representation of such classes of compounds as alcohols, aldehydes, ketones, ethers, acids, esters, fats, and soaps.

20-AV4. "Hydrocarbons and Their Structures." Color. 13 min. Coronet. Bonding properties of carbon; homologous series.

20-AV5. "Organic Chemistry." 40 frames. FOM. Carbon bonding in terms of Kekule and Lewis concepts. The Geneva nomenclature system.

20-AV6. "Organic Chemistry." Color. 40 frames. McGraw-Hill. Basic ideas of organic chemistry.

22-AV7. "Physical Chemistry." Color. 21 min. AT&T. Demonstrations and experiments involving polymerization.

22-AV8. "The Story of Plastics." 40 frames. FOM. Chemistry and uses of.

20-AV9. "Techniques of Organic Chemistry." Color. Four reels; 11, 20, 12, and 19 min. McGraw-Hill. Fractional distillation, melting point determination, determination of molecular weights, extraction, crystallization, elution chromatography, synthesis, etc.

Ideas for Additional Instructional Activities

20-AD1. Bassham, J. A., "The Path of Carbon in Photosynthesis," *Scientific American,* 206: 88–100. June, 1952.

20-AD2. Caldwell, Harrie E., "Discovering Organic Chemistry," *The Science Teacher,* 35: 63–67. November, 1968. Game-type activity involving types of organic compounds.

20-AD3. Caserio, Marjorie C., "Reaction Mechanisms in Organic Chemistry," *Journal of Chemical Education,* 42: 570–576 and 627–632. October and November, 1965.

20-AD4. Clapp, Leallyn B., THE CHEMISTRY OF THE OH GROUP. Englewood Cliffs, New Jersey: Prentice-Hall, 1967. By concentrating on the chemistry of only one group, the author probes rather deeply the relationships between structure and properties.

20-AD5. Clevenger, Sarah, "Flower Pigments," *Scientific American,* 210: 84–88. June, 1964. Chemistry and genetics of.

20-AD6. Greenwood, C. T., and Ann Milne, NATURAL HIGH POLYMERS. Edinburgh: Oliver and Boyd, 1968. Structure, properties, and function of giant molecule substances, especially as they are involved in living systems.

20-AD7. Lanes, Rose M., and Donald G. Lee, "Chromic Acid Oxidation of Alcohols," *Journal of Chemical Education,* 45: 269–270. April, 1968. A simple experiment on reaction rates.

20-AD8. Morgan, Paul W., "The Nylon Rope Trick," *Journal of Chemical Education,* 42–12. January, 1965.

20-AD9. O'Driscoll, Kenneth F., THE NATURE AND CHEMISTRY OF HIGH POLYMERS. New York: Reinhold. 1964. For students who may want to know "less than all" about molecules consisting of more than 100 repeating units.

20-AD10. Park, Roderic B., "Advances in Photosynthesis," *Journal of Chemical Education,* 39: 424–429. August, 1962.

21

BIOCHEMISTRY

The chemistry of living systems is rapidly becoming a highly challenging, importantly exciting branch of chemistry; and rightly so. But this trend poses several complex problems for the high school chemistry course. If we try to capture student interest in biochemical phenomena by pointing out some of the most significant chemistry of living systems, our students will want to pursue their interests, usually, in the laboratory. And they seem to expect to find "answers" to even the biggest physiological or psycho-chemical questions that are just as logical and complete as they are accustomed to in less complex phases of their course.

Increasing attention to the molecular origins of phenomena is built into the newer school biology courses. Many of our students have been introduced to the complexities of the DNA molecule, the intricacies of one or another of the various efforts to describe the release or storage of energy, theories regarding the mechanism of enzyme action, and similar "big" biochemical topics. Obviously, these introductions will have left widely different impressions on different students—impressions which may influence student attitudes when these or similar topics come up in the chemistry course.

Actually, biochemistry is but one of several "speciality" fields that might be given special attention in the introductory general chemistry course. Some teachers, recognizing this fact, may decide to use biochemistry more to introduce students to the idea of special fields in chemistry than to start students to thinking specifically toward biochemistry. On this basis, the instructional activities for this topic should be chosen more for their appropriateness to illustrate types of investigations or large interest areas than for the clarity with which they develop or add to specific principles or concepts.

Many of the topics in biochemistry take us toward the interface between not only several of the branches of chemistry but also between the natural and other sciences, especially the social and political sciences. Of course, chemistry teachers differ widely in their opinions of what we should do about the opportunities or obligations which are generated at these interfaces. Some teachers, for example, feel perfectly within their proper domain when they elucidate the chemistry of a drug, insecticide, explosive, or food additive but they become very uneasy when questioned about the social or ethical implications deriving from specific uses thereof. In contrast, other chemistry teachers encourage their students to discuss the social or ethical aspects of applied chemistry even without elucidating the fundamental chemical principles or concepts that are involved in the preparation of the substances being applied to solve social problems. As always, the inference is that there is a "proper" position somewhere between the extremes of these points of view.

INSTRUCTIONAL ACTIVITIES: STUDENT

21-1. Biochemistry in Styrofoam. Many of our students seem to be overwhelmed by their first glances at the structural formulas of the compounds which play central roles in the chemistry of living systems. Popular magazine articles and museum exhibits seem to glorify the fantastic complexity of key biochemical molecules. Circumstances such as these suggest the use of seemingly "progressive," almost hammy approaches to such reactions as the synthesis of carbohydrates, energy cycles, enzyme action, and similar biochemical reactions. One such approach assumes an adequate supply of foam plastic balls and pipecleaners to simulate atoms and chemical bonds. The example which follows is based on clarification of one version of the glucose-pyruvic acid cycle.

When you go to your lab stations you will find a supply of black, white, red, green, and yellow foam plastic balls to represent carbon, hydrogen, oxygen, nitrogen, and sulfur atoms and short lengths of pipecleaner to represent chemical bonds. Use these models to build whatever molecule or molecules assigned to you by the label at your lab station. Be back at your seats within 15 minutes with your models.

Now, will the two glucose models and the one fructose model come to the front of the room and will one of the glucose molecules rearrange his structure so as to be a duplicate of the fructose molecule. Take your place beside the glucose model. Will the glyceraldehyde model step up? And will one of the fructose molecules now rearrange his model to form two glyceraldehyde models one of which is to remain in position beside the glucose and fructose models. Continue in the proper sequence of reactions until the total energy release cycle has been shown.

For tomorrow, anticipate similar portrayals of such reaction systems as are discussed in your textbook. Pay special attention to the "polymerization" reactions of the glucose unit, the cycle by which energy is released from acetic acid, and the synthesis of fats and proteins. Don't be upset if we ask you to reverse the "action" in some of the episodes.

INSTRUCTIONAL ACTIVITIES: TEACHER

21-T1. Biochemistry in Current Literature. It is easy to agree that the high school chemistry course should bring its students into contact with sources of information in addition to the textbook and laboratory manual. It is less easy to devise instructional activities which involve current literature and news releases—activities which bring our students into touch with valid interpretations of modern research episodes written in language they can understand.

Teachers seem to rely heavily on *Scientific American* as a source of reports of current research written at a level at least our better students can handle with reasonable effort. *Chemistry,* as the title suggests, provides articles written expressly for school chemistry students. Written at a lower level of reading difficulty, *Science World* and *Current Science* include articles dealing with chemistry among the broad range of articles written for junior and senior high school students.

The problem is to squeeze time from busy schedules to read and discuss magazine articles. Too many of us feel cramped for time to teach the fundamental ideas, principles, and concepts of chemistry and our students seem to be under even greater stress for adequate time to prepare their full array of assignments. Furthermore, many students are very skeptical of the worth of assignments involving "book reports" or regularly scheduled attention to "current science" because of unpleasant memories of teachers who used these activities without convincing their students of their appropriateness.

The time lag between the completion of research episodes and the appearance of their reports in school textbooks creates especially appropriate reasons to use current literature when teaching biochemistry, this branch of chemistry being such a rapidly evolving one. The availability of reprints and periodic publication of collections of related articles from *Scientific American* prompts us to feature this publication in an instructional activity devoted to the use of current literature in school chemistry courses.

The Physics and Chemistry of Life, a collection of eighteen articles originally published in *Scientific American,* presents "a synthesis of the state of knowledge of life that is not to be found between the covers of any other book." Almost equally courageous is the editors' statement that "the questions raised here are the kind that can be answered wholly within the disciplines that explain the behavior of nonliving atoms and molecules." And our students' curiosities should surely be pricked when we tell them that the book deals with such big questions as the origin of life; "how complex living material is built up from the simple molecules of nonliving matter; how it is endowed with the properties we associate with life; how living things reproduce themselves and transmit their characteristics to their offspring" with the final chapter describing the "beginning of our understanding of the electrical basis of thought."

Having introduced our students to the book, under ideal conditions we can rely on the writing of its authors to capture and hold their interest and attention. Under less than ideal conditions, we must inject some kind of additional motivation. An obviously gimmicky example is, "When you read the article entitled, 'Enzymes,' by John Pfeiffer, list the terms he uses which also appear in our textbook's discussion of enzymes and their action." Much less mechanical would be the challenge to spin off the design of a laboratory investigation that we might be able to do by studying

Szent-György's report of his research on muscle tissues and the chemistry of muscular contraction. Or the challenge to reconstruct the hypothesis that was tested by Hodgkin and Keynes by using radioactive potassium as a tracer.

Our students often surprise us with their abilities to use supposedly highly technical, totally adult scientific literature. While doing the background research for *Know Your Skin* and *Cancer: The Search for its Origins,* the senior author's students proved to be totally capable of tracking down episodes in chemistry dealing with 3-pentadecylcatechol as the "culprit" in poison ivy, the probable role of tyrosinase in the production of human skin pigmentation, the probability of cause-effect relationships between the incidence of cancer and changes in the secretion rates of androgens and estrogens, or the possibility that cancerous tissues are triggered by excessive action of the energy-liberating enzyme, catalase.

Student success at "making the literature of chemistry their servants" seems to hinge more on the nature of their assignment than on the sophistication of the literature. When told to bring back specific information, many of our students seem to know how to find their way to wherever the answers are. And librarians seem to be especially cordial toward young people nowadays who are anxious to take advantage of available resources.

Audio-Visual Aids

21-AV1. "Biochemistry and Molecular Structure." 20 min. Chems. Role of molecular structure in controlling physiological activities.

21-AV2. "Inside the Cell." Color. Two reels, 45 and 43 min. Norwood Films. Theory of, factors influencing, and laboratory techniques used in studying enzyme action.

21-AV3. "Thread of Life." 29 min. Indiana University. Theorized relationships between nucleic acid, mutations, and viruses.

Ideas for Additional Instructional Activities

21-AD1. Adams, Elijah, "Poisons," *Scientific American,* 210: 76–84. November, 1959. How they produce death from shock, hemorrhage, or incapacitation of some vital organ.

21-AD2. Agranoff, Bernard W., "Memory and Protein Synthesis," *Scientific American,* 216: 115–122. June, 1967. Investigation of memory loss following blockage of manufacture of protein.

21-AD3. Arnon, Daniel I., "The Role of Light in Photosynthesis,"

Scientific American, 203: 105–188. November, 1960. Conversion of light into chemical energy.

21-AD4. Cole, Lamont C., "Rachel Carson's Indictment of the Wide Use of Pesticides," *Scientific American,* 207: 173–180. December, 1962. A review of the book, *Silent Spring.*

21-AD5. Eccles, Sir John, "The Synapse," *Scientific American,* 212: 55–66. January, 1965. Nerve cell transmission of impulses by chemical action.

21-AD6. Frieden, Earl, "The Chemistry of Amphibian Metamorphosis," *Scientific American,* 209: 119–125. August, 1959. Chemistry adaptations accompanying transition from aquatic to amphibian habitats.

21-AD7. Frieden, Earl, "The Enzyme-Substrate Complex," *Scientific American,* 201: 119–125. August, 1959. Interactions between enzymes and substrates.

21-AD8. Hokin, Lowell E. and Mabel R., "The Chemistry of Cell Membranes," *Scientific American,* 213: 78–86. October, 1965. How substances in cell walls transport substances across cell walls.

21-AD9. Lehninger, Albert L., "Energy Transformations in the Cell," *Scientific American,* 202: 102–114. May, 1960. Features enzyme action.

2-AD10. Porter, R. R., "The Structure of Antibodies," *Scientific American,* 217: 81–87. October, 1967.

21-AD11. Rabinowitch, Eugene, and Govindjee, "The Role of Chlorophyll in Photosynthesis," *Scientific American,* 213: 16–17, 74–83. As a catalyst in the transfer of hydrogen atoms.

21-AD12. Robertson, J. David, "The Membrane of the Living Cell," *Scientific American,* 206: 64–72. April, 1962. At the molecular level.

21-AD13. Rose, Anthony H., "New Penicillins," *Scientific American,* 204: 66–71. March, 1961. By rearranging groups of atoms attached to the core of the penicillin molecule.

21-AD14. Stong, C. L., "Experiments With Ptyalin Demonstrate the Basic Principles of Enzyme Action," *Scientific American,* 208: 147–154. January, 1963.

22

DESCRIPTIVE CHEMISTRY

We need go back into the history of chemistry teaching only a dozen or so years to find the high school chemistry course consisting pretty much of descriptions of the occurrence, preparation, properties, and uses of selected elements and compounds. Sheer description overshadowed efforts to identify orderly relationships which might link together large bodies of observations. Of course, there was always the assumption, stated or implied, that the students would weave for themselves the big ideas of chemistry by integrating apparently isolated bits of information.

During most recent years, curricular research has sought an approach to high school chemistry in which the big ideas of chemistry and the thought processes of these big ideas may be nurtured in the minds of students, might be approached in a more head-on fashion. Critics of what they felt to be an overly direct approach to the major principles and concepts of chemistry worried lest students were being asked to comprehend and appreciate ideas for which they lacked the essential direct experience with the precipitates, bubbles, flames and smokes, smells and sounds of laboratory bench chemistry.

Obviously, the two approaches to high school chemistry, "descriptive" versus "seeking orderly relationships" compliment each other. And although the ferment of controversy marked the thinking of many chemistry teachers during the 1950's and 1960's, today's courses are moving toward providing students with the best experiences inherent in the "descriptive" approach while paying much attention to the elaboration of the big ideas, the principles and concepts of chemistry.

Problems in integrating these two approaches seem to be primarily those of perspective. For example, if a student heats potassium chlorate in the presence of manganese dioxide to "prepare oxygen" and remains totally unaware of the bond breakings and bond formings taking place in the total system, the educational values in the exercise are not being exploited effectively. At the same time, unless our students know the excitement of seeing steel wool sparkle into iron oxide in an oxygen environment, we may be asking the improbable when we expect them to sense the elegance we have built into one or another metal model to account for the oxidation of iron.

What seems to be needed are instructional activities which build well traveled bridges between observations experienced at the laboratory bench and the verbal discussions in textbooks and lecture-recitations—discussions devoted to establishing theoretical bases in which isolated observations can be fit into major concepts.

INSTRUCTIONAL ACTIVITIES: STUDENT

22-1. Some Thoughts About the Element, Carbon. Each of the elements which make up the universe, of course, can be interesting to anyone who becomes involved in one way or another with its properties. In so far as arousing interest, however, carbon has many things going for it. It plays a vital role in the chemistry of living systems. When in one of its crystalline forms, diamond, the element evokes ideas of all sorts. And almost any pencil mark, any "carbon" copy, or printed word bears witness to the ubiquity of uses man finds for the element.

Here is an exercise that invites you to produce a bit of elemental carbon that very probably has never been seen as such by any other human being. In fact, if you choose your raw materials wisely, you can produce a bit of carbon that has been locked up in compounds long before the life process ever got started, at least on this planet.

Obtain a chunk of limestone and to play the game of this exercise all the way, you will have to find out where this sample of limestone came from. That is, you will have to establish the geological stratum in the earth's crust that feeds the quarry, outcrops along a road cut, or otherwise allows you and the sample to get together. How to obtain this information is your problem but geologists and earth science teachers should be able to help.

Fit a gas collecting bottle with a one-hole stopper carrying bent glass tubing so you can collect the gas generated in this bottle by the usual water displacement method. Put 10 to 15 grams of the crushed limestone in the bottle, pour in 10 to 15 ml of approximately 2 M hydrochloric acid, insert the stopper, and collect a bottle or two of the generated carbon dioxide. Using crucible tongs, ignite a short length of magnesium ribbon and insert the burning magnesium quickly into a bottle of the carbon dioxide. Supposedly, the burning magnesium breaks the carbon-oxygen bonds in the carbon dioxide, the oxygen combines with the magnesium to produce magnesium oxide, and elemental carbon deposits in tiny globs on the sides and bottom of the bottle.

Collect some of these globs and smear them on a piece of filter paper. If you are a bit philosophically inclined, think about when it was that this carbon, once before, might have existed in elemental form. In fact, you might enjoy the reaction of an English teacher upon reading your essay or theme based on your recreation of the "life story" of this glob or smear of elemental carbon, when and where it became imprisoned in a chunk of limestone, and the bonds you had to sunder to release its atoms and allow them to reassemble. Actually, a more interesting tale could be spun if you select a sample of limestone that contains plant or animal fossils. But such a sample denies you the chance to say that you are the first living system to see or feel the elemental carbon obtained from it. It would be an easy matter, however, for you to claim that you are the first homo sapiens to see this bit of carbon, assuming you select the right sample of fossiliferous limestone.

22-2. The Property of Solubility. In many respects, each element and each compound has a unique set of properties. In some cases uniqueness stems more from the degree to which the element or compound possesses the property, solubility, for example. To investigate this idea, put about 5 ml of 1 M silver nitrate in a test tube. Add 10 ml 1 M sodium sulfate and shake thoroughly. Allow to stand for several minutes in order for the precipitate to settle (and to give you time to record your observations).

After the precipitated silver sulfate has settled, pour off and discard the supernatant liquid without losing any of the precipitate. Now add approximately 8 ml 1 M potassium dichromate and shake thoroughly. Again, allow this system to settle while you are recording your observations. After 4 or 5 minutes, again discard the supernatant liquid and wash the precipitate by adding a few ml of water and shaking. Discard the wash water. Now add approximately 7 ml 1 M sodium chloride solution, shake thoroughly, and set aside to settle while you are recording your observations.

Once more, discard the supernatant liquid and then add about 6 ml of 1 M ammonium sulfide solution. Shake thoroughly and set aside to settle. Record your observations of this system in its final state. Write the equations for the reactions occurring in each of the test tubes. Why were you advised to do the individual steps in the order you did? What would have been the effect, for example, if you had interchanged the final and second steps? If you are not sure of your answer, try this interchange of steps and see what happens.

Looking back through the entire instructional activity, should we say that a substance is *insoluble* or would it be more precise to say that one substance is *less* soluble than another?

INSTRUCTIONAL ACTIVITIES: TEACHER

Experienced teachers accumulate a repertory of "quick and easy" to bring off reactions in which the "preparation, properties, and uses of" selected elements and compounds can be brought before students. Furthermore, these types of instructional activities predominate in the literature of chemistry teaching. The authors present here only a sampling of such activities, albeit many of these are long-standing favorites.

22-T1. The effect of the sun setting through dust-laden atmosphere can be demonstrated by throwing a spot-light through a 700 ml water solution of 5 g sodium thiosulfate to which 5 ml 10 M hydrochloric acid has just been added. As the colloidal sulfur forms, scattered blue light can be seen at a 90° angle from the beam. On the screen or wall, the spot from the beam changes from white through yellow, red, and, finally, it is blacked out completely.

22-T2. Dissolve a few crystals of cobaltous chloride and potassium thiocyanate in a graduate one-third filled with ether. Add an equal volume of water to separate red and blue colors. Now add a few drops of silver nitrate solution to complete the red-white-blue color display.

22-T3. Prepare a tincture of iodine and potassium iodide solution (0.1 M for each solute). Place a drop on a microscope slide and project on a screen with a microprojector. The heat from the lamp speeds up evaporation of the solvent producing beautiful color and crystallization effects.

22-T4. For a simple version of the ammonia fountain, fit a 500 ml flask with a 1-hole rubber stopper carrying a glass tube drawn into a jet inside the flask and attached to a long, heavy rubber tube. Put 5 ml concentrated ammonium hydroxide in the flask, replace the stopper, and heat until ammonia gas comes from the open end of the rubber tube. Place the end of the rubber tube under water and remove the heat from under the ammonia flask.

22-T5. Put equal volumes of 0.5 M sodium carbonate solution into two cylinders. To one, add a few ml 1.0 M sodium acetate to serve as a buffer. Compare reaction rates when 10 ml 2.0 M acetic acid is added to each cylinder.

22-T6. Salt shakers covered with pressure tape are convenient for storing powdered metals or salts that are to be sprinkled into flames to show flame colors, ion colors, or the burning of metals.

22-T7. Small quantities of oxygen can be prepared by the action of enzymes in yeast on 6% solution of hydrogen peroxide, by the surface action of manganese dioxide on 3 or 6% hydrogen peroxide, heating sodium nitrate (*caution*), or the interaction of sodium peroxide and water (*caution*).

22-T8. To demonstrate flame colors, dissolve a small quantity of the salt to be demonstrated in ethanol and ignite in a clean evaporating dish.

22-T9. Lead iodide, when washed and recrystallized from hot water, produces interesting crystals.

22-T10. To show slight solubility of even "insoluble" substances, grind very fine a small amount of glass, put in a test tube with a few ml of distilled water, and add phenolphthalein.

22-T11. Fill two tall cylinders with a saturated solution of sodium bicarbonate. Add a few ml of licorice extract to one. Add a few ml of saturated aluminum sulfate solute to both cylinders to observe the effect of colloidal stabilization.

22-T12. Melt sodium acetate in its own water of crystallization by heating a sample in a small flask in a water bath. Allow the supersaturated solution to cool slowly and with a minimum of mechanical disturbance. Add a crystal of solid sodium acetate to the cool solution.

Audio-Visual Aids

22-AV1. "Ammonia." 14 min. Imperial Chemical Laboratories. (Note: Except where noted, each of the films to follow deals pretty much with the preparation, properties, and uses of the substance or substances indicated in the title.)

22-AV2. "Boron and Aluminum." 44 frames. FOM.

22-AV3. "Bromine Chemistry." Color. Modern Learning Aids.

22-AV4. "Bromine—Element from the Sea." Color. 26 min. Chems.

22-AV5. "The Chemistry of Period 3." 44 frames. FOM.

22-AV6. "The Discovery of the Inert Gases." Color. 19 min. Nederlandse Onderwijs Film, Holland, and the OECD in Association with the International Council for Educational Films.

22-AV7. "The Halogens" and "Preparation and Properties of the Halogens." Color. 11 and 32 min. Coronet.

22-AV8. "High Temperature Research." 20 min. Color. Chems. Properties of substances at high temperatures.

22-AV9. "Hydrogen." Color. 13 min. Coronet.

22-AV10. "Hydrogen—The Fundamental Atom." 40 frames. FOM.

22-AV11. "The Joyful Lunatic." 30 min. Carousel Films. The career of Joseph Priestly.

22-AV12. "Metals of Groups IA and IIA." 44 frames. FOM.

22-AV13. "Nitric Acid." 18 min. Color. Chems. Includes references to Le Chatelier's Principle, K_A, and E°.

22-AV14. "Nitric Acid Compounds and the Nitrogen Cycle." Color. 18 min. Coronet.

22-AV15. "Nitrogen and Ammonia." Color. 16 min. Coronet.

22-AV16. "The Noble Gases." 40 frames. FOM.

22-AV17. "Oxides of Nitrogen." Color. 16 min. Coronet.

22-AV18. "Oxygen." Color. 11 min. Coronet.

22-AV19. "Preparation of Oxygen—Seven Demonstrations." Color. 14 min. Yale University.

22-AV20. "The Sodium Family." Color. 16 min. Coronet.

22-AV21. "The Transition Elements." 42 frames. FOM.

22-AV22. "Vanadium, a Transition Element." Color. 20 min. Chems. Features different oxidation states.

22-AV23. "Zinc Controls Corrosion." Color. 38 min. American Zinc Institute. Includes discussion of fundamental nature of corrosion.

Ideas for Additional Instructional Activities

22-AD1. Alyea, Hubert N., *Journal of Chemical Education,* "Carbon and Its Inorganic Compounds," 45: 325–326, April, 1968; "Nitrogen Chemistry," 45: 567–569, July, 1968; "The Phosphorus Family," 45: 673–674, September, 1968. Demonstrations and/or laboratory exercises.

22-AD2. Frey, John E., "Discovery of the Noble Gases and the Foundations of the Theory of Atomic Structure," *Journal of Chemical Education,* 43: 371–373. July, 1966.

22-AD3. Gelman, Henry, and John J. Eisch, "Lithium," *Scientific American,* 208: 88–102. An outstanding example of descriptive chemistry.

22-AD4. Johnson, Ronald C., *Introductory Descriptive Chemistry,* New York: W. A. Benjamin, 1966. Uses the properties of a few of the non-metallic elements to convey a bit of the "excitement which is always present on the frontiers of science."

22-AD5. Jolly, William L., *The Chemistry of the Non-Metals,* Englewood Cliffs, New Jersey: Prentice-Hall, 1966. Structural, kinetic, and thermodynamic properties of eighteen elements together with the noble gases.

22-AD6. Lavoisier, Antoine-Laurent, *Elements of Chemistry,* Translated by Robert Kerr and with a new introduction by Douglas McKie, New York: Dover, 1965. The facsimile format adds authenticity to this unabridged republication of the translation first published in 1790.

22-AD7. Rochow, Eugene G., *Organometallic Chemistry,* New York: Reinhold, 1964. Preparation, properties, and uses of those compounds in which organic groups are linked directly to metal atoms, and only to them.

22-AD8. Sanders, R. T., "Principles of Halogen Chemistry," *Journal of Chemical Education,* 41: 361–366. July, 1964. Penultimate shell, van der Waals forces, bond energies, halogenating power, reducing power, electronegativity.

22-AD9. Selig, Henry, "The Chemistry of the Noble Gases," *Scientific American,* 208: 66–77. May, 1964.

22-AD10. Staff, "Gallium," *Chemistry,* 40: 16–18. June, 1967. Discovery, properties, and applications.

23

ANALYSIS AND SYNTHESIS

Analysis and synthesis procedures provide good opportunities to introduce students to the practical and productive aspects of chemistry. Instructional activities involving these procedures also encourage students to bring their knowledge to bear on actual performance. And when attempting to solve specific analytical problems or synthesize "standard" samples, students see how several principles and concepts can interact within the same system or reaction. Additional educational values in the minds of some teachers can be obtained from analysis and synthesis assignments, especially in the laboratory, because the successful pursuit of such assignments requires well thought through plans and disciplined attention to precise procedures. No teacher-imposed discipline is quite as effective as the chagrin that comes with discarding the wrong filtrate, allowing the reaction temperature to go too high, or finding oneself with two similar but unlabelled flasks containing totally different substances.

Students need instructional activities which give them laboratory sense. It is always amazing how many students know of a "test" for starch but how few know of a "test" for iodine. Many students seem to be thrown off course when an instructor glibly suggests substituting a reagent for the one specified in a laboratory exercise. At the other extreme, some students glibly alter suggested procedures or quantities of reactants and are surprised to find their anticipated results do not materialize.

It is easy for students to get caught up by the challenge to identify their assigned "unknown" or to add to their collection of "standard samples" and lose sight of the knowledge we hoped would be strengthened through application. Well timed discussions diminish the probability of this happening.

INSTRUCTIONAL ACTIVITIES: STUDENT

23-1. Qualitative Analysis. Show a person a chunk of something and quite often the first question asked is, "What's in it?" But this is a difficult question for even a well equipped chemist to answer. For one reason, there could be nearly a hundred different elements in the sample or thousands of different compounds. Before a chemist could go to work on the question he needs to restate the question in such form as: Does this material contain silver, sulfur, calcium, gold, or whatever other elements he suspects might be involved. Actually it would take a brave (or well paid) chemist to tackle the problem of analyzing a sample of a substance that might be suspected of containing any or all of the earth's elements. Perhaps you will see why after carrying out this instructional activity.

On the demonstration table is a solution containing all of the following ions: Ag^{+}, Pb^{+2}, Fe^{+3}, Co^{+2}, Ni^{+2}, Al^{+3}, Cr^{+3}, and Zn^{+2}. In other containers you will also find solutions obtained by dissolving various objects in nitric acid, for example, a foreign coin, a piece of cast off jewelry, a mineral specimen, or a piece of the metal trim from an automobile. With your lab partner, obtain 10 ml portions of the "known" and one of the "unknown" solutions. Run both solutions through the following analytical procedure using the "known" solution to acquaint you with the evidence that you will need to prove the presence or absence of each of the "known" elements in the "unknown" solution.

The strategy underlying this system relies on separating the ions of the "known" solution by taking advantage of differences in solubilities and by subjecting the ions to reactions which produce observations which are unique for each ion in the mixture. For example, in Procedure 1, by adding chloride ions to the solution, all positive ions present in the solution which form relatively insoluble chlorides are thus separated from the solution in the form of a precipitate. In this way, the silver, mercury(I), and lead(II) ions are converted to chlorides and the mixture of chlorides can be subjected to a series of reactions whereby the three metals can be identified.

The remaining solution is then treated with NH_3 solution and a group of relatively insoluble hydroxides precipitate from the solution. Iron(III), chromium(III), and aluminum are in this group. The nickel, cobalt, and zinc form soluble ammonia complexes and remain in solution. Again, specific reactions are used to identify each of the metals in both of these groups. See Figure 23-1 for a flow sheet which shows the complete procedure.

If centrifuges are available, use 13 $\times$ 100 mm test tubes and separate precipitates by centrifuging. Without centrifuges, you can fall back on the usual filtering process. In either case, all glassware should be clean and steps taken to avoid undue contamination. This exercise demands careful labeling of containers and recording all significant observations.

Specific directions for this exercise appear in many laboratory manuals. To illustrate the procedure, for the first step, place approximately 1 ml of the "known" and the "unknown" solutions in two clean test tubes. Add 6 to 8 drops of 6 M HCl. Place the test tubes in a beaker of cold water for a minute or so. Stir gently, centrifuge, and pour off the supernatant liquids into other test tubes, label, and set aside for Procedure 5. Wash the precipitate by adding 1 ml cold water containing a drop of HCl. Centrifuge and add this liquid to the solution being saved for Procedure 5. Treat the precipitate by Procedure 2. Etc.

You will appreciate this exercise much more and understand better the underlying strategy if you write equations for each reaction in the procedure.

23-2. Synthesis of Serpent's Eggs. Chemical reactions which have surprising results are always interesting. All kinds of fireworks, for example, have a special kind of satisfaction. But too many young people in years past have crippled or even killed themselves by messing around with home-made fireworks to permit unnecessarily dangerous reactions to be brought off in school laboratory programs. In fact, one investigator who looked into fifty "accidents" found that 67 people were injured including 13 people killed.

Here is one reaction somewhat of the fireworks type that can be brought off safely. The principal material, mercuric thiocyanate, as

Flow Sheet for Qualitative Analysis Laboratory Exercise

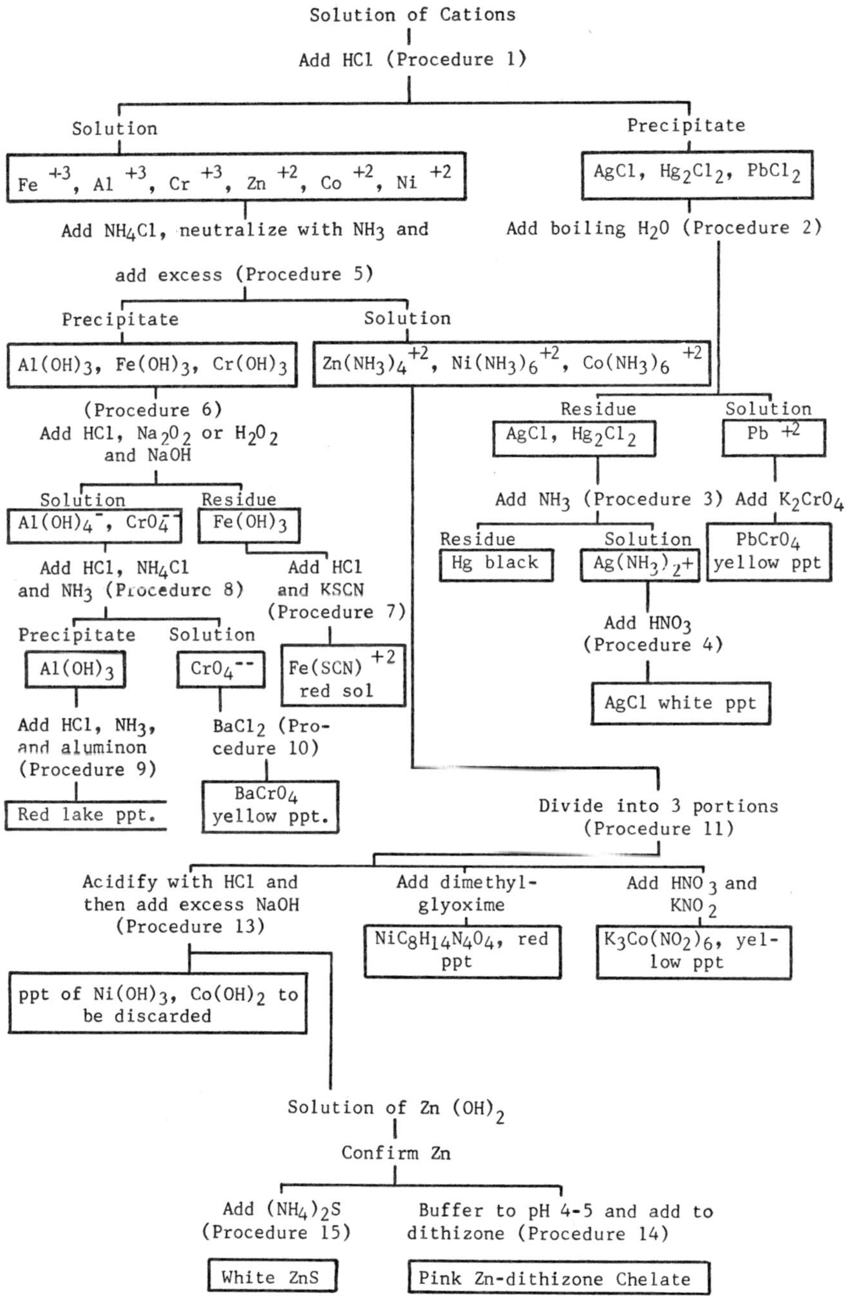

Figure 23-1

well as the substances needed to synthesize it, however, are *very toxic.*

Using no more water than necessary, dissolve about 2 g of potassium thiocyanate in a small beaker. In a second 50 ml beaker, prepare a similar saturated solution of mercury(II) nitrate using about 3.5 g of the solid. Verify stoichiometrically that these weights provide approximately two SCN for each Hg ion. Before combining the solutions, however, add a few drops of 1 M ferric chloride solution to the mercury solution. The ferric ions will let you know when all mercury ions have been tied up as mercuric thiocyanate because they will then form the characteristic red ferric thiocyano ion.

With constant stirring, add the KSCN solution to the mercury solution until a faint red color persists. Let stand for 30 minutes or overnight, then filter, wash the precipitate several times with cold water, and then dry in a warm place. To avoid spontaneous ignition, however, use no heat. When dry, transfer the precipitate to a small mortar and work into a paste by adding a few drops of water containing mucilage and a few grains of potassium nitrate. Work into cones about one-fourth inch in diameter and one-half inch tall. When dry, ignite a cone on an asbestos mat.

Audio-Visual Aids

23-AV1. "Chemical Instruments." 14 min. University of Southern California. Colorimeter, fluorimeter and other instruments for somewhat advanced analytical procedures.

23-AV2. "Fermentation Products, Chemistry Laboratory Demonstration." 14 min. Purdue University. Preparation and purification of ethanol.

23-AV3. "Paper Chromatography." Color. 14 min. Thorne Films. Simple and refined methods. R_f values determined for chlorophyl and amino acid extracts.

23-AV4. "Principles of Chromatography." Color. 20 min. McGraw-Hill. One and two-dimension paper chromatography.

23-AV5. "A Research Problem: The Inert Gas Compounds." Color. 20 min. Chems. Illustrates research procedures through the preparation and identification of krypton difluoride.

23-AV6. "Small Scale Organic Preparations." 17 min. Imperial Chemical Industries. Manipulative techniques in small scale operations illustrated by the preparation of ethyl bromide and the conversion of nitrobenzene to aniline.

23-AV7. "Synthesis of a Compound." Color. 13 min. Coronet. Synthesis of copper sulfide to illustrate calculation of empirical formula.

23-AV8. "Synthesis of an Organic Compound." Color. 22 min. Chems. Uses 2-butanone to illustrate synthesis, purification, and identification.

Ideas for Additional Instructional Activities

23-AD1. Anderson, Larry B., and Charles N. Reilley, "Teaching Electroanalytical Chemistry," *Journal of Chemical Education,* 44: 9–16. January, 1967. A broad array of electrochemical concepts and principles.

23-AD2. Carleton, Robert H., "Are Vegetable Dyes Pure Substances or Are They Mixtures of Different Dyes?" and "To Determine Which Acids Are Contained in Orange Juice," *The Science Teacher,* 22: 37–38. February, 1955. Paper chromatography.

23-AD3. Carter, Giles F., "Ancient Coins as Records of the Past," *Chemistry,* 39: 14–19. November, 1966. Applying modern analytical methods to an interesting material.

23-AD4. Chernick, Sidney S., "Use of Commercial Diagnostic Reagents," *The Science Teacher,* 27: 27. December, 1960. Glucose utilization by yeast, invertase in yeast, and protein digestion.

23-AD5. Connes, Pierre, "How Light is Analyzed," *Scientific American,* 219: 72–83. September, 1968. Spectroscopy.

23-AD6. Keller, Roy A., "Gas Chromatography," *Scientific American,* 205: 58–67. October, 1961. Separating the components of complex mixtures.

23-AD7. Kuhn, Edward T., "Construction and Use of a Spectroscope in High School Chemistry," *The Science Teacher,* 31: 62–63. April, 1964.

23-AD8. Lauren, Paul M., "Chalk Chromatography, Three Dimensional Separation," *The Science Teacher,* 33: 56–57. January, 1966.

23-AD9. Mueller, J. Robert, and Brian M. Davis, "Thin Layer Chromatography," *Chemistry,* 39: 29–31. November, 1966. Introduction to.

23-AD10. Otani, Theodore T., "Two-Dimensional Paper Chromatography," *The Science Teacher,* 27: 29–31. October, 1960. Applications in biochemistry.

23-AD11. Phillips, J. P., "Sixteenth Century Tests on Assaying. Origins of Quantitative Analysis," *The Journal of Chemical Education,* 42: 393–394. July, 1965. History and techniques of assaying.

23-AD12. Pollak, Michael E., "Determining Bromine in Aqueous Solutions by Diffusion Through a Gel," *Chemistry,* 40: 29. September, 1967. Based on diffusion coefficient.

23-AD13. Poonia, N. S., "Paper Chromatography and Spot Tests: a Useful Combination of Techniques," *Journal of Chemical Education,* 43: 423. August, 1966. See also 44: 477–478. August, 1967. Separation of closely related inorganic ions.

23-AD14. Ritchie, A. S., "A Paper Chromatographic Scheme for the Identification of Metallic Ions," *Journal of Chemical Education,* 38: 400–405. August, 1961.

23-AD15. Racusen, David, "Investigation of Unknowns by Paper Chromatography," *Journal of Chemical Education,* 39: 484–486. September, 1962.

23-AD16. Solmssen, Ulrich V., "The Chemist and New Drugs," *Chemistry,* 40: 22–27. April, 1967. Pharmaceutical research.

23-AD17. Stong, C. L., "An Amateur Uses Paper Chromatography to Separate the Constituents of a Mixture," *Scientific American,* 205: 162–172. July, 1961.

23-AD18. Stong, C. L., "Analysis of Blood Serum Proteins by Means of Zone Electrophoresis," *Scientific American,* 206: 171–180. June, 1962.

23-AD19. Stong, C. L., "A Spectrophotometer Is Built for Less Than $75." *Scientific American,* 218: 140–146. May, 1968. And a good bit of skill.

23-AD20. Wahl, W. H., and H. H. Kramer, "Neutron Activation Analysis," *Scientific American,* 216: 68–72. April, 1967. Determining concentrations of elements in amounts less than one part per billion by making them radioactive.

23-AD21. Yohe, G. R., "Coal," *Chemistry,* 40: 19–22. January, 1967. Applications of analytical techniques to an important commodity.

24

NUCLEAR CHEMISTRY

From a somewhat arbitrary point of view, the phenomena of chemistry are limited to those accompanying the rearrangements of atoms from molecule to molecule—rearrangements which do not involve changes in the sub-atomic particles within the nucleus. Nuclear phenomena fall into the domain of physics. The boundaries between the domains of chemistry and physics become confused, however, when alpha, beta, or gamma rays or neutron beams affect the bondings between atoms within molecules. And when the ejection or capture of nuclear fragments affects the atomic number of an atom, the chemistry of the atom is altered abruptly.

A discussion of the transuranium elements invites attention to nuclear phenomena, especially when describing their "man-made" origins. In fact, the episode of the "man-made" elements suggests instructional activities designed to show how the abstractness of atomic theory led to the reality of new forms of matter. And the men who are involved in the "man-made" term provide stellar examples of the methods, strategy, and spirit of science in action.

Man's ability to release and control nuclear energy has had and promises to continue to have great significance to many aspects of our society. Some teachers feel obligated to mull over with their students the problems which nuclear energy exploitation has created at the interface between the natural and social sciences. It is a challenge to create a classroom atmosphere in which everyone's hopes and fears can be aired with maximum objectivity and minimum sensationalism or bias. But again, some teachers feel they must include an effort to meet this challenge concurrently with the clarification of the physical and chemical phenomena involving sub-atomic particles.

INSTRUCTIONAL ACTIVITIES: TEACHER

24-T1. Governmental Subsidy of Nuclear Versus Fossil Fuel Resources—A Student Debate. Although the debate format is used rarely, certain topics arise in the school chemistry course which invite an intelligent and disciplined airing of opinions on controversial issues. An example is the question of investing public funds in the development of nuclear energy versus the improved management or exploitation of petroleum and coal resources.

If properly used, the debate format encourages students to marshall a body of facts and bring them to bear on a topic. And facts dealing with the world supply of petroleum, for example, versus the annual consumption or the economic feasibility of converting public power stations to nuclear reactor power sources threaten to become increasingly relevant.

The debate format also invites our students to examine published points of view for evidence of bias or rhetorical hanky-panky, the identification of which is, in itself, a valuable talent. Actually, the student debater tends to provide glaring examples of faulty reasoning and not always is the class environment of the sort we need to make sure the reasoning is adequately analyzed. Other weaknesses arise from the general habit of students not to take seriously the information or points of view expressed by other students. Keeping in mind the values inherent in school-wide colleague relations, it may be appropriate to invite the school's debate coach to assist in setting up and conducting a debate on the future of nuclear power versus power sources fueled by conventional fuels.

Nuclear science is not the only source of controversial issues likely to confront the chemistry teacher. We may find increasing need for formats which enable us to handle sensitive issues, a need that invites us to add the debate and similar disciplined discussion formats to our repertories.

Audio-Visual Aids

24-AV1. "Beyond Uranium." 29 min. Indiana University. Discovery of six transuranium elements.

24-AV2. "Conquest of the Atom." Color. 22 min. Realist Film Unit in Cooperation with the British National Committee for Visual Aids in Education. Nuclear fission and the control of nuclear energy.

24-AV3. "Discovery of Radioactivity." Color. 15 min. Institut Fuer und Bild., Germany, and OECD in Association with the International Council for Educational Films. Historical survey.

24-AV4. "Madame Curie." 23 min. Teaching Films Custodians. Separation of radium from pitchblende.

24-AV5. "Making Elements." 29 min. Indiana University. Features strategy and equipment used in the discovery of element 101.

24-AV6. "Modern Alchemy." Color. 29 min. Indiana University. The discovery of "missing" and transuranium elements.

24-AV7. "Radiation and Matter." 44 min. USAEC. Interactions of alpha, beta, and gamma radiations and matter.

24-AV8. "Radioactive Series." 9 min. McGraw-Hill. Decay of U-238 to stable lead.

24-AV9. "The Transuranium Elements." Color. 20 min. Chems. Production, separation by ion exchange, and proof of identity of selected transuranium elements.

Ideas for Additional Instructional Activities

24-AD1. Alexander, Peter, "Radiation-Imitating Chemicals," *Scientific American,* 202: 99–108. January, 1960. Including use in cancer therapy.

24-AD2. Anders, Oswald U., "The Place of Isotopes in the Periodic Table, The 50th Anniversary of the Fajans-Soddy Displacement Laws," *Journal of Chemical Education,* 41: 522–525. October, 1964. Includes decay series, displacement laws, and isolation of protactinium.

24-AD3. Bodash, Lawrence, "How the 'Newer Alchemy' Was Received," *Scientific American,* 215: 89–95. August, 1966. History of the Rutherford-Soddy theory involving the transmutation of elements by radioactive decay.

24-AD4. Gay, Helen, "Nuclear Control of Cell," *Scientific American,* 202: 126–136. January, 1960. Transfer of materials from nucleus to cytoplasm studied by tracers.

24-AD5. Grisham, Larry R., "Separation of Strontium in Radioactive Dating of Rocks," *Chemistry,* 40: 28–30. June, 1967. Based on use of nitric-dithizone process.

24-AD6. Kamen, Martin D., "The Early History of Carbon-14," *Journal of Chemical Education,* 40: 234–242. June, 1967. When and how carbon-14 became identified.

24-AD7. Platzman, Robert L., "What is Ionizing Radiation?," *Scientific*

American, 201: 74–84. September, 1959. X-rays, alpha, beta, and gamma rays.

24-AD8. Quinney, Paul R., and Howard A. Swartz, "Demonstration Experiments With Radioisotopes," *Chemistry,* 40: 49–53. July-August, 1967. Involving, for example, diffusion, equilibrium, glucose metabolism.

24-AD9. Seaborg, Glenn T., "Some Recollections of Early Nuclear Age Chemistry," *Journal of Chemical Education,* 45: 278–289. May, 1968. Developments involving the periodic chart since before World War II.

24-AD10. Seaborg, Glenn T., and Arnold R. Fritsch, "The Synthetic Elements," *Scientific American,* 208: 68–78. April, 1963. Elements 102 and 103 are featured.

24-AD11. Sulcoski, John W., "Paper Radiochromatography," *Chemistry,* 40: 45–48. July-August 1967. Separation of Po, Bi, and Pa.

24-AD12. Towne, Jack C., and Bonnie M. Gatti, "Demonstrations of Radioautography and Paper Chromatography Using a Polaroid Camera," *Journal of Chemical Education,* 40: 243–244. May, 1963. Features separation of two carbon-14 labeled amino acids.

24-AD13. Weinberg, Alvin M., "Breeder Reactors," *Scientific American,* 202: 82–94. January, 1960. Reactors that produce more fuel than they consume.

25

GEOCHEMISTRY

The term "geochemistry" has been generated by interest in chemical phenomena occurring on a global scale. Of course, in light of man's achievements in space during recent years, the term should take on the additional meaning to include lunar and, possibly, planetary phenomena beyond those associated with the planet Earth. Although fundamentally broad in scope, geochemists find their projects in finite samples of the earth's (or moon's) crust, the earth's atmosphere and its waters.

Teachers who seek integrating themes turn to geochemical phenomena. As illustrated so effectively by the team effort to describe as accurately as possible the rock samples returned from the moon, nature is not compartmentalized into departments of chemistry, physics, astronomy, geology, biology, or mathematics. Nor is the history of the universe published in separate subject matter textbooks.

Thoughts believed to be good occur when students sense that the changing condition of the earth reflects actions just as orderly, just as predictable as those which accompany the crystallization of fused materials in their test tubes, the deposition of precipitates from reactions in solutions, or the interactions of photons in low pressure gas systems. Something believed to be good occurs when our students sense that the state of man's environment is basically predictable and, hence, manageable. Obviously, to create and administer instructional activities which nurture such highly intangible objectives calls for considerable inventiveness and creativity.

Openendedness has been cited often as one of the chief attributes of good instructional activities for high school chemistry courses. The authors feel it is quite appropriate to close this book by saying that this same attribute, especially as it involves such traits and capabilities as inventiveness and creativity, exists in each of the activities we have attempted to describe. If it is good for students to sense the openendedness in their learning activities, it is good for teachers to sense the openendedness in their instructional activities.

INSTRUCTIONAL ACTIVITIES: STUDENT

25-1. Pure Salt from a Prehistoric Ocean. Obtain a supply of commercial rock salt. The chances are good that this material has been mined from a deposit left somewhere in the earth's crust—a deposit marking the evaporation of a prehistoric ocean or large lake. Perhaps you have visited the Great Salt Lake in Utah. Here, in modern times, salt is being deposited by the relentless evaporation of a lake. Were the continent to undergo new submergences and emergences as it has during past geological ages, the salt beds of the Great Salt Lake could be covered with layers of sediment and, eventually, packed into new rock salt strata.

If you set out to obtain pure sodium chloride from rock salt, you can assume that it might contain as impurities any or all of the substances that are dissolved in seawater. Either you will have to arrange to remove each impurity from the sodium chloride or you will have to arrange to remove the sodium chloride from all of the impurities. The suggested procedure to follow, in a way, combines these two strategies.

Weigh out approximately 30 g of rock salt. Use precise weighing if you choose to match your wits and hands against trying to extract a maximum yield of pure salt from your raw material. Dissolve the rock salt in 90 ml of hot water. If some of the salt fails to dissolve, either you are too impatient or the rock salt contained an insoluble impurity, some kind of silicates, for example. Filter and retain the filter paper so it can be weighed when dry to see what weight of impurity was removed by this first step.

To the filtrate, add 1.0 g of anhydrous sodium or potassium

carbonate dissolved in 5 ml lukewarm water. Hopefully, this step will remove all metallic elements which form insoluble carbonates, particularly, calcium and magnesium. Before filtering, allow the system to settle for several minutes and then add a few drops of the carbonate solution to see if the precipitation of carbonate forming precipitates is complete. If it is not, add more of the carbonate solution. Filter and weigh the filter paper when dry to see what weight of impurities was removed by this step.

Assuming you now have a concentrated solution of reasonably pure sodium chloride, the strategy of the next step is to remove the sodium chloride from the remaining impurities. The idea is to take advantage of the common ion principle by generating hydrogen chloride gas and allowing it to saturate the sodium chloride solution. Hopefully, the mass action of the newly arriving chloride ions will drive the original sodium chloride out of soluton causing it to crystallize and, thus, enabling it to be filtered from the system.

See the Figure 9-1 for a suggested set-up for generating hydrogen chloride and dissolving it in the salt solution. Put approximately 90 g of sodium chloride in the generating flask and use 11.6 M hydrochloric acid in the wash bottle. Locate the set-up under the *hood* before filling the dropping funnel with 17.8 M sulfuric acid. Do not begin dropping the sulfuric acid on the sodium chloride until your instructor approves the procedure and then do so slowly. The total reaction including the precipitation of purified sodium chloride in the original rock salt solution should take 20 minutes or so. Do not attempt to disassemble the set-up when finished without help from your instructor.

Filter the recrystallized sodium chloride from the newly formed solution of hydrochloric acid, dry, and weigh. Examine a sample of the crystals under a lens. Does the regularity of crystal form and size suggest you have a "pure" sample of sodium chloride?

Ideas for Additional Instructional Activities

25-AD1. Garstang, Michael, "Fueling of the Air-Sea System," *School Science and Mathematics,* 67: 707–716. November, 1967. Energy exchanges on a global scale.

25-AD2. Lyman, John, "Marine Geochemistry," *The Science Teacher,* 34: 27–30. May, 1967.

25-AD3. Marshall, L. C., and Nadine F. George, "The Evolution of Oxygen and Nitrogen in the Earth's Atmosphere: With Notes on the

Evolution of Life on Earth, and the Martian Atmosphere." *School Science and Mathematics,* 69: 13–30. January, 1969. Treated from the perspective of many branches of science.

25-AD4. Mason, Brian, "Organic Matter From Space," *Scientific American,* 208: 43–49. March, 1963. Using known knowledge of carbon compounds to hypothesize the origin of meteor associated "organic" matter.

25-AD5. Moutvic, J. C., "Oceanography in the High School Curriculum," *The Science Teacher,* 34: 55–57. May, 1967.

25-AD6. Starkman, Ernest S., "Chemical Pollution from Transportation Vehicles," *School Science and Mathematics,* 68: 477–487. June, 1968. How motor vehicles contribute to air pollution.

25-AD7. Van Landingham, John W., "Inorganic Nutrients in the Sea," *Journal of Chemical Education,* 40: 272–276. May, 1963. Chemical, physical, and biological problems of the marine environment.

APPENDIX A

ADDRESSES FOR OBTAINING AUDIO-VISUAL AIDS

AT&T
American Telephone and Telegraph Co.
195 Broadway
New York, N. Y. 10007

American Zinc Institute, Inc.
292 Madison Ave.
New York, N. Y. 10017

Carousel Films, Inc.
1501 Broadway
New York, N. Y. 10036

CHEMS (Chem Study)
Modern Learning Aids
3 East 59th Street
New York, N. Y. 10022

Chem. Lab. Tech.
Association Instructional Films
600 Madison Ave.
New York, N. Y. 10022

CORONET
65 East South Water St.
Chicago, Illinois 60601

Disraeli Films
Box 343, Cooper Station
New York, N. Y. 10003

Edmund Scientific Co.
300 Edscorp Building
Barrington, New Jersey 08007

Encyclopaedia Britannica
425 North Michigan Avenue
Chicago, Illinois 60611

Film Associates
11559 Santa Monica Blvd.
Los Angeles, California 90025

Films, Inc.
5625 Hollywood Blvd.
Hollywood, California 90028

FOM
Filmstrip-of-the-Month Clubs, Inc.
355 Lexington Ave.
New York, N. Y. 10017

Imperial Chemical Industries, Ltd.
488 Madison Avenue
New York, N. Y. 10022

Indiana University
Audio-Visual Center
Bloomington, Indiana 47401

International Film Bureau
International Council for Educational Films
332 South Michigan Ave.
Chicago, Illinois 60604

The Jam Handy Organization
2821 E. Grand Blvd.
Detroit, Michigan 48202

Long Films
Charles J. Long
931 W. Elsmere Place
San Antonia, Texas 78201

MCA, Manufacturing Chemists' Association
1825 Connecticut Ave., N. W.
Washington, D. C. 20009

McGraw-Hill Co.
330 West 42nd St.
New York, N. Y. 10036

Modern Learning Aids
3 East 54th Street
New York, N. Y. 10022

NITL, National Instructional Television
Library
10 Columbus Circle
New York, N. Y. 10019

Norwood Films
926 New Jersey Ave.
Washington, D. C. 20001

Purdue University
Audio-Visual Center
Lafayette, Indiana 47907

Rothacker, Inc.
Time and Life Building
New York, N. Y. 10020

Society for Visual Education
1345 Diversey Parkway
Chicago, Illinois 60614

State University of Iowa
Audio-Visual Center
Ames, Iowa 50010

Thorne Films
1229 University Ave.
Boulder, Colorado 80302

University of Minnesota
A-V Education Service
Wesbrook Hall
Minneapolis, Minnesota 55455

University of Southern California
Department of Cinema
Film Distribution Division
University Park
Los Angeles, California 90007

The Welch Scientific Company
7300 North Linder Avenue
Skokie, Illinois 60076

Yale University
Educational Motion Picture Production Department
New Haven, Connecticut 06520

BIBLIOGRAPHY

Some General References

ASIMOV, ISAAC, *A Short History of Chemistry.* New York: Doubleday and Co., 1965.

BENFEY, O. THEODOR, *Classics in the Theory of Chemical Combination.* New York: Dover Publications, 1963.

DRUMMOND, AINSLEE, JR., *Chemical Reactions.* Columbus, Ohio: American Education Publications, 1965. A brief overview of representative principles and concepts of chemistry.

GARRETT, ALFRED, *The Flash of Genius.* Princeton, New Jersey: D. Van-Nostrand, 1963. Discoveries in the field of chemistry and physics.

LAPP, RALPH AND THE EDITORS OF *Life*, "Matter." New York: Time, Inc., 1963. Highly illustrated overview of selected fundamentals of chemistry.

LEICESTER, HENRY, *The Historical Background of Chemistry.* New York: John Wiley and Sons, 1956.

PARTINGTON, J., *A Short History of Chemistry.* New York: Harper, 1957.

SCIENTIFIC AMERICAN, EDITORS OF, *New Chemistry.* New York: Simon and Shuster, 1957. Selected episodes.

WOODBURN, JOHN H., *Opportunities in the Chemical Sciences.* New York: Vocational Guidance Manuals, 1971. Intended to help chemistry teachers with their career guidance efforts.

INDEX TO AUTHORS

A

Adams, Elijah, 193
Agranoff, Bernard W., 193
Alexander, Classen, 132
Alexander, Leroy E., 144
Alexander, Peter, 211
Allen, J. A., 171
Alyea, Hubert N., 40, 72, 95, 123, 132, 145, 158, 179, 201
Anders, Oswald U., 211
Anderson, Larry B., 207
Andrew, H. G., 95
Armstrong, George T., 179
Arnon, Daniel I., 193
Asimov, Isaac, 218
Astin, Allen V., 31
Atkinson, R. C., 115
Austin, Leonard G., 179
Azaroff, L. V., 151

B

Bachman, G. B., 132
Bacon, E. K., 95
Bacon, James E., 171
Bacon, W., 171
Barnard, Walther M., 64
Barrett, Richard L., 95
Barrow, Gordon M., 171
Bassham, J. A., 187
Bauman, Robert P., 171
Benfey, Theodor, 151, 218
Bent, Henry A., 151
Berman, Robert H., 72
Berry, Stephen R., 158
Bishop, J. A., 95
Boblick, John M., 113
Bodash, Laurence, 211
Bragg, Sir Lawrence, 158
Breedlove, C. H., Jr., 115

C

Caldwell, Harrie E., 187
Campbell, J. Arthur, 165
Carleton, Robert H., 207
Carmody, Walter R., 64, 151
Carnell, P. H., 172
Carter, Giles F., 207
Caserio, Marjorie C., 187
Chandler, E. E., 72
Chernick, Sidney S., 207
Child, W. C., Jr., 64
Choppin, Gregory R., 145
Clak, John N., 172
Clapp, Leallyn B., 116, 187
Clevenger, Sarah, 187
Clever, H. Lawrence, 145

Coffin, C. C., 72
Cohen, Irwin, 64
Cole, Lamont C., 194
Connes, Pierre, 207
Conway, B. E., 133
Cooper, M. J. R., 72
Copley, George Novello, 54, 64
Cottrell, A. H., 151

D

Darnowski, Vincent S., 115
Davenport, D. A., 73
Davidson, D. L., 179
Davis, Brian M., 207
Davis, J. C., 95
Davis, John D., 73
Demo, J. J., 82
Dence, Joseph B., 165
Derjaguin, Boris V., 145
DeVries, Dale, 48
Dickieson, Samuel C., 138
Dillard, Clyde R., 133
Dinga, Gustav P., 24
Dingledy, David, 64
Donnelly, L. R., 81
Dorfman, M. K., 96, 115
Drummond, Ainslee, 218
Dutton, Frederick B., 40, 81, 123, 132, 145

E

Eccles, Sir John, 194
Eddy, R. D., 95
Edwards, J. O., 116
Edwards, Thomas H., 165
Ehrenreich, Henry, 133
Eichlinger, Jack W., 158
Eisch, John J., 201
Ellis, George, 172
Ellis, R. H., 115, 133
Eyring, Edward M., 165
Eyring, Henry, 165

F

Fabishak, Vernon L., 54, 123
Farrell, J. K., 132
Feigl, F., 81
Feldman, L., 133
Ferguson, Harold W., 124
Fernelius, W. C., 152
Ferreira, Ricardo, 158
Fowles, G., 133
Frey, John E., 201
Frieden, Earl, 194
Furst, A., 81

G

Gamow, George, 158
Garrett, Alfred B., 158, 218
Garstang, Michael, 215
Gay, Helen, 211
Gayer, Karl H., 124
Gehman, W. G., 151
Gelman, Henry, 201
Gentile, Philip S., 158
George, Nadine F., 215
Gillespie, R. J., 115
Gilman, Alfred J., 116
Gorman, M., 124
Glasstone, Samuel, 73
Gray, Harry B., 165
Greenwood, C. T., 187
Grisham, Larry R., 211
Guggenheim, E. A., 64

H

Haas, Rudy, 124
Habashi, Fathi, 151
Haggin, Joseph H. S., 95, 179

Hall, Marie Boas, 73
Hammer, A. J., 115
Hammond, George S., 165
Hanson, A. L., 95
Hausman, Eugene A., 151
Henry, Paul S., 64
Hill, Derek, 179
Himes, D., 73
Hirschfelder, Joseph O., 158
Hoare, J. P., 96
Hokin, Lowell E., 194
Hokin, Mabel R., 194
Hollahan, J. R., 73
Howald, Reed A., 138
Huang, Suei-Rong, 54

J

Johnson, Ronald C., 210
Johnston, Helen, 54
Jolly, William L., 201
Jones, H. R., 82
Jones, Jerry L., 96

K

Kamen, Martin D., 211
Kammeyer, Patty Hall, 133
Karp, Stewart, 54
Kauffman, George B., 96
Keller, Roy A., 207
King, L. Carroll, 48, 139
Klemm, L. H., 96
Kobe, K. A., 96
Kokes, R. J., 96, 115
Kramer, H. H., 208
Krieder, James, 48
Kuhn, A. T., 133
Kuhn, Edward T., 207

L

Lambert, Frank L., 158
Lanes, Rose M., 187
Lange, N. A., 96
Lapp, Ralph, 218
Lapp, Walter S., 115
Lauren, Paul M., 124, 133, 207
Lavoisier, Antoine-Laurent, 201
Lee, Donald G., 187
Lee, J. H., 96
Lee, Shiu, 64
Lehninger, Albert L., 194
Leicester, Henry, 218
LeRoy, L. F., 73
Lingane, J. J., 179
Linner, E. R., 115
Livingstone, R., 179
Lodge, James P., Jr., 73
Lansdale, Kathleen, 145
Lucasse, W. W., 96, 124, 180
Lyman, John, 215

M

Mahan, B. H., 82, 180
Marshall, L. C., 215
Martin, D. R., 124
Mason, Brian, 216
Masterton, W. L., 82
Mathia, T., 96, 115
McClenon, Robert, 96
McGuire, W. S., 117
Meloan, Clifton E., 124
Miller, J. B., 124
Miller, J. G., 96, 180
Milne, Ann, 187
Morgan, Paul W., 187
Morris, Kelso B., 172
Morris, L. H., 116
Moss, Stephen J., 179
Moutvic, J. C., 216
Mueller, J. Robert, 207

N

Nechamkin, Howard, 96, 139
Neidig, H. A., 31, 65, 133

Neilsen, E. K., 48
Newton, T. W., 124
Nicholson, Douglas G., 165
Nowako, Andrew V., 158

O

Obourn, Ellsworth S., 40
O'Driscoll, Kenneth F., 188
Oppenhuizen, J., 97
Otani, Theodore T., 207
Owen, Brewster, 180

P

Palit, S. R., 97
Palmer, W. G., 97
Parish, C., 133
Park, Roderic B., 188
Partington, J., 218
Pattison, D. B., 124, 180
Pauling, Linus, 82, 151
Pearson, A. David, 145
Peiser, H. Steffen, 145
Peterson, B. H., 97
Phillips, J. P., 207
Platzman, Robert L., 211
Pollak, Michael E., 207
Poonia, N. S., 208
Porter, George, 180
Porter, R. R., 194
Poutsma, Marvin, 48

R

Rabinowitch, Eugene, 194
Rabinowitch, Gavindjee, 194
Racusen, David, 208
Ramette, R. W., 64
Rauch, R. B., 133
Reber, P., 73
Reilley, Charles N., 207
Ries, Herman E., 48
Ritchie, A. S., 208
Robertson, J. David, 194
Robey, R. F., 152
Rochow, Eugene G., 201
Ronneberg, C. E., 116
Rose, Anthony H., 194
Rossini, F. D., 82

S

Salomon, Mark, 133
Sanders, R. T., 201
Sanderson, R. T., 82, 133, 139
Schaff, John F., 172
Scholes, S. R., 64
Schubert, Leo, 32
Seeger, W., 82
Selig, Henry, 201
Shan, N. M., 116
Shen, Samuel Y., 54, 116
Siegel, Bernard, 180
Sienko, M. J., 172
Sime, R. J., 151
Smith, Cyril S., 40
Snyder, Asa E., 97
Solmssen, Ulrich V., 208
Stakely, William N., 72
Starkman, Ernest S., 216
State, Harold M., 116
Stock, John T., 116
Stong, C. L., 97, 116, 133, 158, 194, 208
Strong, Laurence E., 48, 116
Strong, Robert L., 179
Swann, Peter R., 152
Szabadvary, Ferenc, 116

T

Taube, Henry, 165
Theiman, Theo, 64

Thiessen, G. W., 65
Torgesen, John L., 145
Ts'ao, Ching-Pan, 116
Turner, A. Mason, 116

V

Van Landingham, John W., 216
Van Tamelen, Eugene E., 54
Veguilla-Berdecia, Luis Antonia, 139
Verhoek, Frank H., 152
Videen, Tom, 166
Visser, M., 97

W

Wahl, W. H., 208
Walker, Ruth A., 54
Wartsburg, Arthur F., 73
Weaver, Elbert C., 82
Webb, Valerie J., 152
Westmeyer, Paul, 172
Wherry, Edgar T., 115
Wiersma, Jack, 151
Wilchers, Edward, 54
Wilhoit, Randolph C., 180
Williams, Buford, 158
Williams, L. Pearce, 133
Wilmshurst, J. K., 139
Wilson, H. A., 115
Wilson, Mitchell, 180
Wolthuis, Enno, 48, 97
Woodburn, John H., 40, 48, 218

Y

Yamanashi, Bill S., 158
Yingling, R. T., 65
Yohe, G. R., 208
Youden, W. J., 32

Z

Zeldin, Martel, 145
Ziman, John, 180
Zuder, W. F., 117
Zuelke, R. W., 116
Zuffanti, S., 117
Zwolenik, James J., 166

GENERAL INDEX

A

Acid anhydrides, 99
Acid-base indicators, 101, 103, 114
Acid-base theories, 114
  electronic, 117
Acids, 98
  strengths, 115
Activated complex, 179
Activation energy, 175
Actual vs theoretical yields, 58, 60
Alkalis, 98
Allotropism, 144
Alloys, 147, 152
Ammonia fountain, 92, 199
Amphoterism, 100
Analogs, 104
Analysis, 202
Analysis of data, 22
Analytical balance, 29
Anode, 127
Antibodies, 194
Arithmetic, 55
Assaying, 207
Assumptions, 142
Atomic and molecular models, 47
Atomic numbers, 78, 156
Atomic orbitals, 77
Atomic structure, 48
Atomic theory, 201
Atomic weights, 42, 65
  determination, 63, 65
  scale, 54
Autocatalysis, 164
Avogadro's hypothesis, 57, 63, 66, 72
Avogadro's number, 44, 47, 48, 64, 132, 133

B

Balances, 31
  sensitivity, 26
Balancing equations, 49, 54, 123, 124
"Barking dog" demonstration, 39
Bases, 98
  strengths, 115
Biochemistry, 167, 189, 193
Bohr, Niels, 155
Boiling point, 141, 151
  elevation, 91
Bonding, 134, 165, 177
  energies, 138
  geometry, 151
  theory, 134
  types, 143
Book reports, 192

Boyle, Robert, 67, 73
Boyle's law, 66
Breeder reactors, 212
Bronsted-Lowry, 98, 103, 104
Brownian movement, 92
Buffers, 105, 199
Burettes, 31

C

Caloric, 180
Calorimetry, 76, 82, 176, 179, 180
Carbohydrates, 190
Carbon bonding, 187
Catalysis, 51, 131, 162
Catephoresis, 92
Cathode, 127
Cause-effect, 140
Celsius, 68
Cell membrane, 194
Charge symmetry, 143, 145
Charles' law, 66, 73
Chemical change, 23, 33, 34, 37, 175
Chemistry, 191
Chromatography, chalk, 207
  elution, 187
  gas, 207
  paper, 206, 207
  thin layer, 207
Circular logic, 140
Classification of oxides, 82
Coefficients, 49
Coins, 207
Collegative properties, 83
Colloidal state, 92, 198, 199
Colors, patriotic, 38
Combining capacities, 52
Combining weights, 64
Combustion, 37, 166
Common ion effect, 170, 215
Compounds, 18
Computer language, 111
Computerized lesson, 106, 112, 115
Computational skills, 55
Conceptual definition, 135
Concomitance, 140
Conjugate acid, 103
Conjugate base, 103
Conservation of atoms, 51, 59
Consumer education, 33
Coordinate covalent bond, 138
Coordination, 96
Corrosion, 132, 148, 151, 201
Co-solvency, 97
Covalency, 136, 143
Cryoscopy, 115
Crystal garden, 38
Crystals, 61, 95, 97, 124, 144, 145, 150, 151, 152, 158, 170, 175, 180, 187, 199, 215
Curie, Madame, 211
Current literature, 191
Current Science, 191

D

Davy, Humphrey, 133
De Broglie, Louis, 155
Deduction, 136, 142
Definite composition, 48
Delta-G, 82, 177
Delta-H, 50, 82, 177, 179
Delta-S, 82, 177
Density, 27
  of air, 29
Desalting water, 97
Descriptive chemistry, 195
Designing experiments, 89, 105, 120, 154
Diagnostic reagents, 207

Diffusion, 73, 208
Dimensional analysis, 32
Dirac, P. A. M., 155
Discharge, gaseous, 47
"Discovery" vs reinforcement, 20
Disorder, laws of, 180
Displacement, 81
Dissociation, 132
DNA, 189
Drugs, 208

E

E°, 121, 164
Economic problems, 91
Effusion, molecular, 73
Einstein, Albert, 155
Electrical conductivity, 85, 101
Electrical interactions, 92
Electroanalytical chemistry, 207
Electrochemistry, 118, 124, 125, 131, 132, 177
Electrode potential difference, 177
Electrodeposition, 132
Electrolysis, 92, 133
Electromotive force, 132, 179
Electronegativity, 76, 81, 127, 136, 139, 143, 151, 186, 201
Electron energy levels, 47
Electron transfer, 118
Electronic theory of acids and bases, 117
Electrophoresis, 208
Electroplating, 128
Electropositive, 136
Elements, 18, 23
  combining capacities, 24
  names and symbols, 24, 50
  states of, 18, 53
Empirical formula, 64, 207
Endothermic, 50
Energy cycles, 190
Energy level, 77, 157
Enthalpy, 95, 115, 124
Entropy, 173, 177
Enzymes, 190, 192, 193, 194, 199
Equations, 53
  determining, 37, 96, 167, 169, 171, 179
Equilivalent weight, 48, 82, 116
Exclusion principle, 158
Exothermic, 176

F

Families, 78, 81
Faraday's electrochemical laws, 130, 132
Fermentation, 206
Fire safety, 29, 30
Fireworks, 204
Flame colors, 199
Flower pigments, 187
Formulas, 49, 53
  determining, 61, 63
Fractional distillation, 187
Free energy, 171, 173
Freezing point depression, 91, 95, 96
Fuel cells, 179
Functional groups, 186
Fused substances, 86

G

Games, 24, 51, 187
Gas constant, 68
Gas laws, 66, 71
Gas reactions, first and second order, 72
Gases, properties, 72, 171

Geochemistry, 213, 215
Goudsmit, S. A., 155
Graduated cylinders, 31
Graham's law, 66, 69, 73
Gravimetric titration, 115
Guide sheets, 120, 137

H

Half-cell reactions, 127
Half-reaction equations, 118
Halogenation, 201
Handbook of Chemistry and Physics, 17, 60, 184
Heat additive law, 176
Heat content, 173
Heisenberg, W., 155
Hess's law, 176, 179
History of chemistry, 45
Homologous series, 182, 187
Hydration, 90
Hydronium ion, 145
Hydrogen bonding, 143, 152, 186
Hydrogen fountain, 72
Hydrogen ion, 100
Hydrophylics, 96
Hydroxide ion, 100
Hygroscopic, 61
Hypodermic syringes, 73
Hypotheses, 88, 142, 154

I

Ideal gas, 66
Indicators, 101, 103, 114, 115, 116
Inert gas atom, 143
Inert gas compounds, 206
Instructional quiz format, 136, 142, 156
Instruments, 206
  precision, 25
Intermolecular forces, 145
Invisible inks, 38
Ion bridge, 127
Ionic bonding, 136, 138, 143
  radius, 76
Ionization, 92, 171
  energy, 76, 81, 136, 151, 186
  potential, 81
  radiations, 211
Isomerism, 184
Isotopes, 211

K

K_A, 103, 104, 168, 172
Kelvin, 68
Kinetic-molecular theory, 72
Kinetics, 158
K_{sp}, 90

L

Laboratory, role of, 20
  attitudes and habits, 22
Language of chemistry, 52
Lasers, 145
Latent heat, 80
Lattice energy, 82
Lavoisier, Antoine-Laurent, 210
Lead storage battery, 171
LeChatelier's Principle, 171
Light sensitive materials, 47
Literature, use of, 192

M

Macromolecular bonds, 138
Magic number, 8, 75
Matter and energy, 30
Mass action, 215
Mass vs weight, 25
  volume and density, 27

Measurement, 31
statistical treatment, 31
Mechanism of reaction, 160
Melting point, 187
Mendelyeev, 74
Mental model, 142, 145
Metals, 81, 123, 146, 151, 175
Metallic oxides, 82
Metallic bond, 150, 151
Metallurgy, 147
Metamorphosis, 194
Metaphors, 104, 159
Metathesis, 115
Mixtures, 18
Models, 52, 77, 78, 125, 136, 145, 152, 162, 190
Molarity, 63, 101
Mole, 42, 44, 46, 64
Mole volume, 69
Molecular architecture, 139
Molecular crystal, 143
Molecular motions, 165
Molecular symmetry, 151
Molecular weights, 187
of solutes, 95
determinations, 97
estimating, 84
Mono-basic acids, 103, 111
Monomolecular films, 48
Moseley, G-J., 75
Multi-basic acids, 103

N

Names of substances, 17
Nerve impulse, 194
Network solid, 143
Neutralization, 60, 100, 115
Neutron activation analysis, 208
Nobel, Alfred, 123
Normality, 101
Nomenclature, 187
Non-metals, 81, 123, 146, 201
Nuclear chemistry, 209
Nuclear fission, 211
Nuclear fuel, 210
Nucleic acids, 193
Nylon rope trick, 187

O

Observations, 78, 136, 142, 154
Oceanography, 216
Openmindedness, 214
Operational definitions, 135
Orbitals, 48, 77, 138, 155, 156, 158
Orderly relationships, 195
Organometallics, 201
Orientation in space, 155
Oxidation, 118, 123, 127, 187, 200
numbers, 118, 183
potential, 50, 123, 127
-reduction, 64, 118, 124, 165
series, 182
Oxides, 114
Oxidizing agent, 127

P

Packing, 151
Paper radiochromatography, 212
Parables, 104
Particle nature of matter, 41
Pauli, W., 155
Penicillin, 194
Percentage weight composition, 64
Periodic table, 74
Periodicity, 75, 82
Pesticides, 194
Pfeiffer, John, 192
pH, 104, 106, 115, 169

Phenon concept, 180
Photochemistry, 48, 158, 166
Photoelectric effect, 154
Photolysis, 158, 166
Photon, 157
Photosynthesis, 187, 188, 193, 194
Physical change, 34, 37, 140
Pipettes, 30
Planck, Max, 154
Planck's constant, 155, 157
Plastics, 187
Poisons, 193
Pollution, 216
Polarity, 83, 90, 91, 139, 151
Polymerization, 187, 191
Potentiometry, 116, 179
Precipitates, 36, 96
Prefixes and suffixes, 51
Pre-lab instruction sheet, 112
Priestley, Joseph, 200
Principal quantum number, 155
Programmed instruction, 64, 106, 113
Projects, 90
Properties of substances, 17, 19
Proteins, 193
Proton donors, 98
Puzzling situations, 28, 36, 37, 79, 168
Pycnometer, 28

Q

Quantitative analysis, 132, 203
Quantum numbers, 155
Quantum theory, 47, 133, 153, 156, 158
Questions, asking, 23
  types, 159
Quiz sheet format, 129, 130

R

Radioactivity, 211
  dating, 211
  series, 211
Radioautography, 212
Radiochromatography, 212
Radioisotopes, 212
Randomness, 165, 173
Raoult's law, 95
Rate determining step, 161
Ratio and proportion, 56
Reaction mechanism, 160, 165, 187
  rate, 151, 162, 164
Reducing agent, 127
Reduction, 82, 123, 127
"Reinforcement" vs "discovery," 20
Relative atomic weights, 42, 56
Relative vs actual weights, 42
Relativistic explanations, 141
Rumford, Count, 180
Rutherford-Soddy theory, 211

S

Safety in the laboratory, 30
Salts, 100, 114
Schlieren, 89, 92
Schrodinger, E., 155
Science World, 191
Scientific American, 191
Seawater, 214
Serpent's eggs, 204
Sigma bond, 136
"Silent" demonstration, 78
Silica gel, 96, 97, 124
Silver tree demonstration, 151
Size of atoms, 81
Social implications, 190
  problems, 91

Soil testing, 116
Solar energy, 168
Solubility, 37, 60, 83, 197
  curve, 95, 96
  gases in liquids, 92
  product, 90, 95
Solutions, 83
  energy interactions, 87, 88, 92
Solvent action, 97
Space, organic matter, 216
Specific gravity, 28, 30
Specific heat, 174
Spectrophotometry, 158, 208
Spectroscopy, 47, 207
Spin, 155
Spot-reactions, 81, 208
Standard solutions, 64, 115
Statistical treatment, 31
  confidence limits, 32
  dimensional analysis, 32
Stoichiometry, 55
STP, 63
Stress corrosion failure, 152
Styrofoam models, 52
Sub-atomic particles, 210
Sublimation, 151
Subscripts, 49
Sub-shell, 155
Substances, properties, 17, 40
Suffixes and prefixes, 51
Surface tension, 95
Supersaturated, 200
Symbols, 49
Synthesis, 202

T

Temperature vs heat content, 174
Theoretical vs actual yields, 58, 60
Theory, 136, 154
Thermocouples, 179
Thermodynamics and equilibria, 82, 171
Thermometers, 31
Thermistors, 180
Thomson, J. J., 155
Titration, 101, 112, 115
Titration curves, 106
Tolerances, 31
Tracers, 211
Transport reactions, 165
Transuranium elements, 209, 210
Trends in properties, 76
Tyndall effect, 92

U

Uhlenbeck, G. E., 155

V

Valence, 77, 81, 136, 158
  electrons, 150, 176
Van der Waals forces, 143, 201
Vapor pressure, 70, 72, 95
Vegetable dyes, 207
Vibrations of molecules, 165
Viscosity, 95
Volta's pile, 133
Volumetric flasks, 31, 64

W

Water pollution, 91
Water vapor pressure, 63, 69
Wave length, 157
Weight and volume problems, 63
Weight vs mass, 25, 30
Werner, Alfred, 96

X

X-rays, 144